AMERICAN PHILOSOPHICAL QUARTERLY
MONOGRAPH SERIES

AMERICAN PHILOSOPHICAL QUARTERLY
MONOGRAPH SERIES

Edited by NICHOLAS RESCHER

STUDIES IN THE PHILOSOPHY OF SCIENCE

Essays by:

Peter Achinstein
Keith Lehrer
Lawrence Sklar
Mario Bunge
Bernard R. Grunstra

Simon Blackburn
Stephen Spielman
Joseph Agassi
D. H. Mellor
Michael Anthony Slote

Q175
S933
1969

Monograph No. 3   Oxford, 1969

PUBLISHED BY BASIL BLACKWELL
WITH THE COOPERATION OF THE UNIVERSITY OF PITTSBURGH

250934

© *in this collection Basil Blackwell 1969*
631 11470 X

Library of Congress Catalog
Card No.: 78-89653

PRINTED IN ENGLAND
by C. Tinling & Co. Ltd., Liverpool, London and Prescot

CONTENTS

	Editor's Preface NICHOLAS RESCHER	7
I.	Explanation PETER ACHINSTEIN	9
II.	Theoretical Terms and Inductive Inference KEITH LEHRER	30
III.	The Conventionality of Geometry LAWRENCE SKLAR	42
IV.	What Are Physical Theories About? MARIO BUNGE	61
V.	The Plausibility of the Entrenchment Concept BERNARD R. GRUNSTRA	100
VI.	Goodman's Paradox SIMON BLACKBURN	128
VII.	Assuming, Ascertaining, and Inductive Probability STEPHEN SPIELMAN	143
VIII.	Popper on Learning From Experience JOSEPH AGASSI	162
IX.	Physics and Furniture D. H. MELLOR	171
X.	Religion, Science, and the Extraordinary MICHAEL ANTHONY SLOTE	188
	Index of Names	207

EDITOR'S PREFACE

This is the third volume in the Monograph Series inaugurated by the *American Philosophical Quarterly* in 1967. The ten papers presented here deal with a wide range of topics in the philosophy of science and represent a variety of approaches. All, however, serve as indicators of the intensity and acumen with which investigations in this domain are nowadays pursued. The *American Philosophical Quarterly* is most grateful to the learned contributors for authorizing the inclusion of their essays in this collection.

The editor acknowledges with thanks the collaboration of his wife in helping to edit the work and to see it through the press.

Nicholas Rescher
Pittsburgh
November, 1968

I
Explanation
PETER ACHINSTEIN

PHILOSOPHERS who write about explanation usually distinguish what they call its "logical" aspects from its "pragmatic" ones. Having done so they concentrate solely on the former, and as a result tend to create a one-sided and, I believe, inaccurate picture. I want to propose an analysis that will have certain advantages over such accounts: it will cover a larger group of explanations; it will spell out the contextual nature of explanation; it will cite criteria for evaluating explanations and indicate how their applicability depends on the situations in which explanations are offered. Since this analysis appeals to the concept of understanding, I plan to say something about this as well.

1. "Explain"

In one sense, everyone would agree that the authors of the Bible explain the origin of man; in another sense, many would disagree. The former sense might be paraphrased by saying that the authors of the Bible attempt to explain, the latter by saying that they correctly or satisfactorily explain. In the first three sections I shall be concerned with the former sense, and so will use the expression "attempt to explain." My question is what does it mean to say that someone would attempt to explain something q by citing something E.

What someone would cite in attempting to explain q depends in part on the sort of person to whom he is explaining. How I would attempt to explain the rainbow, Boyle's law, or the concept of entropy to children in the third grade will differ in important ways from how I would do this in a college physics course. Furthermore, when I attempt to explain q I may cite several things or just one. If several, one of them may be the most central, and I may be willing to call it my explanation of q. I shall refer to what is cited as E, where this may be one thing or several. We have, then, the following schema:

(1) A would attempt to explain q to those in situation S by citing E, or by citing a number of things of which E is the most central.

What can be substituted for the various schematic letters? For A (the explainor), persons and groups of persons. For E (what is cited in explaining), one or several events, facts, phenomena, states of affairs, propositions (including things such as laws, theories, and books, which are, or are composed of, propositions). For S (the situation), information about the knowledge and particular concerns of persons to whom A is attempting to explain q. For q (the explained), many things, e.g., a . . . (fact, state of affairs, event, phenomenon, concept, proposition, theory, book, poem, etc.), the . . . of . . . (structure of the atom, meaning of this poem), as well as many sorts of questions in *oratio obliqua* form, e.g., why . . . (that happened), how . . . (a steam engine works), what . . . (occurred, the difference is between a baryon and a lepton), how possibly . . . (that could have occurred), and so forth. It will simplify the presentation if we can suppose that any q that is explained is, or is transformed into, a question or set of questions in *oratio obliqua* form. Instead of "A would attempt to explain this phenomenon" we can say "A would attempt to explain how (or why) this phenomenon occurs," or "A would attempt to explain what the nature of this phenomenon is." In what follows, Q will be used to designate a question and q the *oratio obliqua* form of that question.

Invoking the concept of understanding, schema (I) can, I suggest, be expressed as follows:

(II) A would attempt to render q understandable to those in situation S by citing E, or a number of things of which E is the most central, as providing what A believes is or might be a correct answer to Q.

I shall defer until later an analysis of understanding. Several preliminary points must be made regarding (II).

1. Often people attempt to explain things by providing what they believe to be a correct answer. But sometimes their attempt is made in the spirit of exploration. If a physicist is in the midst of developing a nuclear theory, he might attempt to explain why certain nuclear phenomena occur by providing not what he definitely believes is a correct answer but what he believes might be correct. Still, it may be objected, can't A attempt to explain q to those in S by deliberately providing what he believes to be an incorrect answer to Q? There are conflicting temptations here. One is to agree, admitting that it is possible for A to attempt to explain something incorrectly. The other is to disagree, saying that A is not really attempting to explain q to those in S, but rather attempting to mislead them into thinking that what he cites provides a correct explanation of q, or attempting to indicate how, according to some false doctrine, q is explained, or something of the like. These conflicting temptations reflect what I

take to be different and conflicting uses of the expression "attempt to explain," and I reply to the above objection by saying that schemas (I) and (II) are concerned only with the latter use of this expression.

2. I am using "understandable" in such a way that if q is understandable to those in S, then those in S actually understand q. So we can also say that if A is attempting to explain q to those in S by citing E, then he is attempting to make those in S understand q by citing E as providing what he believes is or might be a correct answer to Q. The fact that someone is not attempting to render q understandable to those in S in such a way counts decisively against saying that he is attempting to explain it to them, and the fact that he is attempting to render q understandable to those in S in this way counts decisively in favor of saying that he is attempting to explain q to them.

This might be denied. It might be claimed that A could cite E in order to render q understandable to those in S without necessarily attempting to explain q to them. Suppose, e.g., I know that the statement "God is love" is so causally efficacious with you that the mere uttering of it will cause you to understand anything, and in particular why atoms emit only discrete radiation. If so, I might say "God is love" thereby attempting to make you understand why atoms emit only discrete radiation, even though by saying this I would be making no attempt to explain anything to you. But this in no way shows that (II) inadequately expresses (I) for I do not believe that "God is love" does or might provide a correct answer to the question "Why do atoms emit only discrete radiation?", and this is required by (II).

The converse might also be rejected. It might be denied that attempting to explain q to those in S requires attempting to render q understandable to those in S. Suppose a physicist introduces the concept of entropy to students in a third grade class and proceeds in the same way he does in his college thermodynamics class, knowing full well that the third graders will not be able to understand what he is saying, and he is not attempting to make them understand this. Isn't he, nevertheless, attempting to explain to them what the concept of entropy means? He is attempting to cite to those in S what he believes is a correct answer to Q, and this may make us want to say that he is attempting to explain q to those in S. But then we must add that he is treating those in situation S as if they were in situation S'; he is attempting to explain q to those in S under the supposition that they are not in S. So we might also say, with appropriate emphasis, that he is not really attempting to explain q *to those in S*. Either way, we cannot say, at least not without qualification, that our physicist is attempting to explain q to those in S. What necessitates the qualification is precisely the fact that he is not attempting to render q understandable to those in S.

3. Usually if A is attempting to explain q to those in S then it is implicit that those in S do not already understand it. But sometimes they do, and A may know this. A may attempt to explain the concept of entropy to his teachers who already understand it perfectly well. However, if he does this then there is a sense in which it can be said he is treating them as if they did not understand this concept. Accordingly, in (II), when I speak of attempting to render q understandable to those in situation S, I mean to those who do not, or (in the appropriate sense) can be treated as if they do not, understand q. We can say that to attempt to explain q to those in S by citing E is to attempt to bring it about that if those in S do not understand q, citing E will make them understand q, and it is to attempt to do this by providing what the explainor believes is or might be a correct answer to Q.

4. To attempt to explain q to those in S one need not attempt to render q understandable in all respects, i.e., so that no further questions could be raised concerning it. Rendering q understandable to those in S involves providing an answer to Q. This in turn may involve providing an answer to other questions that might be raised in S by those who do not understand q. It does not involve providing an answer to every question that might be raised by anyone at all who does not understand q. What questions may need to be answered will vary with the knowledge and concerns of those in S and with the sort of explanation the explainor thinks appropriate under the circumstances (more about this in Sect. IV).

5. I have spoken about attempting to explain. What about successfully explaining? If we say that A successfully explained q to those in S by citing E, this would mean that A succeeded in rendering q understandable to those in S by citing E as providing what A believes is or might be a correct answer to Q, i.e., by citing E as providing such an answer, A succeeded in getting those in S to understand q.

II. Explanation

Can schema (I), understood in the sense of (II), be used to define "explanation"? To begin with, "explanation" has an act-content ambiguity. In referring to A's explanation we may be referring either to his act of explaining or to what explains, according to A. In the former sense we can speak of A's explanation as something that took so much time and was given at a certain place. In the latter sense we cannot, because we are referring not to A's act of explaining but to what was cited in the performance of this act. I am concerned with the latter sense. "Explanation" has a second ambiguity. If I describe something as being an explanation of q, I may mean that it is a good or satisfactory explanation, or my description may carry no such implication. I am concerned with the latter sense of "explanation," the sense in which bad as well as good explanations are classifiable as explanations.

Schema (I) contains the formula: "A would attempt to explain q to those in situation S by citing E." The variable 'E' ranges over such things as events, facts, phenomena, states of affairs, and propositions. These are items that are cited when someone attempts to explain something. These are also items that can be said to be or to provide explanations. We may say of a mine cave-in (an event), or of the fact that the mine caved in, or of the weakened conditions of the walls of the mine that led to the cave-in (a state of affairs), or of the proposition expressed by the sentence "the mine caved in," that it is or provides an explanation of what caused the death of the miners. If we say that an E *is* an explanation of q, we are using the "is" of classification, not identity. So the formula we want to consider is this: "E is classifiable as being or providing an explanation of q." How shall this formula be defined? Appealing to schema (I) we might be tempted to propose the following definition:

> E is classifiable as being or providing an explanation of q if and only if there is some A and some situation S such that schema (I) is satisfied, i.e., such that A would attempt to explain q to those in S by citing E, or by citing a number of things of which E is the most central.

There are difficulties with this calling for a contextual relativization in the definition. If "There is some A" means "There now exists some A," then the definition precludes explanations that would have been given by past A's, or explanations that would have been given by no actual A in the past, present, or future, but by imagined A's. Suppose we ask: How many explanations of the distribution of galaxies are there? Shall we say two—the steady state and the big bang? Shall we include those offered in the past but completely rejected today? Can we count one my six year old son invented, or one that hypothetical persons might have offered? The answer is that there is no answer independent of the context in which such questions are raised. Whether E is classifiable as an explanation depends in part on the context of classification C in which there is reference to the type of person who would attempt to explain by citing E. The contemporary cosmologist may say there are two explanations, since what he is counting as explanations are only E's actually proposed by leading scientists of the day. Some historians of science may count E's proposed by scientists influential in the 19th and 20th centuries. Still others might be counting E's that no actual person would have proposed but that it is plausible to imagine someone as being willing to propose. Accordingly, I suggest the need to relativize the definition of explanation to the context of classification C.

Whether A would attempt to explain q by citing E depends, as I have said, on the situation S, i.e., on the knowledge and concerns of those to whom he would explain. So the definition of explanation can also be relativized to S. I think that the notion of *someone* attempting to explain something *to someone else* is central for a definition of explanation. There are other uses of the concept of attempting to explain but these are derivative. Thus, textbooks in science are said to attempt to explain and may do so without including information about the knowledge and concerns of those who might read them. But this would not be true unless it is the authors of these texts who are attempting to explain. And the authors consider, in at least a general sort of way, the background and training of their potential readers as well as the sorts of concerns and puzzles such persons are likely to have. They attempt to explain for persons with certain knowledge and interests. S need not always be describable in a precise way, nor need the class of persons referred to have the same background and interests in order that A be said to be attempting to explain q to them.

With this in mind let me propose the following definition:

> (III) Given C, the context of classification, E is classifiable as being or providing an explanation of q, within situation S, if and only if there is (was, might have been) some (type of) A who would attempt to explain q to those in S by citing E or by citing a number of things of which E is the most central. Whether we choose "is," "was," or "might have been," and what type of A is allowed (e.g., leading scientist A) depends on C.

Does this open the floodgates for explanations? In one sense, yes. Given many E's and q's that to us seem quite unrelated, we can at least imagine an A and a situation S such that A would attempt to explain q to those in S by citing E. But this would make E an explanation only relative to that situation and that context of classification. It would not necessarily make it an explanation given a context in which we are considering what E's leading contemporary scientists would cite in explaining q and actual situations in which they would do so. My point is that we need a definition of "explanation" that will permit such flexibility in what can and cannot be counted as an explanation. Sometimes our criterion is narrow, sometimes not.

III. UNDERSTANDING

Since "explanation" is defined by reference to "explain," and this in turn by reference to "understanding," I will need to consider the latter. To do so it will be useful to refer to some work of Sylvain Bromberger.[1] Let us begin with a pair of examples analogous to ones Bromberger employs. In the first example, I am studying Bradley's *Appearance and Reality* and come to the sentence: "The Absolute is each appearance, and is all, but it is not any one as such." I do not know what this means; but worse, any hypothesis I can conceive regarding its meaning is precluded by what I know or believe, or is incorrect, or both. (Under "conceive" Bromberger includes activities such as imagine, conjure up, invent, and remember.[2]) In the second example, I am studying the same book and come to the same sentence, whose meaning I do not know. But this time I can at least conceive of hypotheses regarding its meaning none of which is precluded by what I know or believe, or one of which is correct, or both. In each of these cases I do not know what the sentence means. However, according to Bromberger, in the first case, but not the second, I would be said not to understand what it means.

More generally, Bromberger's account can be expressed as follows. Let Q be some question and q the *oratio obliqua* form of Q. Any statement of the form "A does not understand q" (e.g., "A does not understand what the above sentence from Bradley means") "may be used to report or describe situations in which *either* (1) none of the answers that the person mentioned at 'A' can conceive is an answer that that person can accept—and this includes situations in which the person spoken about can conceive of the right answer, but cannot accept it in the light of his or her other beliefs—*or* (2) none of the answers that the person mentioned at 'A' can conceive is the right answer—and this includes situations in which that person can conceive of one or more answers that he or she can consistently accept. These statements are thus marked by an ambiguity, and demand contextual clues that indicate whose conditions on the answer are at

---

[1] "An Approach to Explanation" in vol. II of *Analytical Philosophy*, ed. by R. J. Butler (Oxford, 1965), pp. 72–105.
[2] *Ibid.*, p. 82.

play, the speaker's or those of the person spoken about."[3] To extend this account to "A does not understand q" where q is not a question in *oratio obliqua* form (e.g., understanding phenomena, events, facts), Bromberger suggests that the q is always replaceable by a suitable question or set of questions in this form.

This analysis contains some valuable ideas, but there are also some difficulties, as illustrated by the following examples.

(i) Suppose John, who listened to Mary, thinks she said "I love you" when she really said "isle of view." We might then conclude that John does not understand what Mary said, even though neither of Bromberger's conditions (1) or (2) holds; i.e., even though (1) John can conceive of an answer to the question "What did Mary say?" that he can accept (though it is not correct), and (2) John can even conceive of (imagine, conjure up, invent) the right answer (though he cannot accept it).

(ii) Suppose John does not know what Mary said on a certain occasion because he was not present and did not read or hear about it. Suppose further that none of the answers that John can conceive to the question "What did Mary say?" is an answer that he can accept or is an answer that is correct. Then, according to Bromberger's analysis, we must say that John does not understand what Mary said. But this sounds odd. If John was not present to hear Mary and did not read or hear about what she said, the more natural description is simply that John does not know what Mary said, not that he does not understand it.

(iii) Suppose John knows that a certain prisoner escaped by sliding underneath the door of his cell. He can accept this as a correct answer to the question "How did the prisoner escape?" because, let us say, he actually saw the prisoner escape in this manner. But he finds it very puzzling, very implausible, in the sense that it is so inconsonant with his other beliefs. If so, we might say that although he knows a correct answer to the question "How did the prisoner escape?", he does not understand how the prisoner escaped. On Bromberger's analysis, however, we are debarred from saying this, since neither condition (1) nor (2) is satisfied. More generally, someone may have a very strong, even decisive, reason r for believing

---

[3] *Ibid.*, p. 83. If condition 1 obtains and A also believes that Q admits of a right answer, then A is in what Bromberger calls a p-predicament with regard to Q (p. 82). If condition 2 obtains and Q in fact admits of a right answer, then A is in what Bromberger calls a b-predicament with regard to Q (p. 90). Bromberger goes on to propose a definition of "A explained q to B" which uses the concepts of p- and b-predicaments and is different from my account in important respects; e.g., it is concerned only with good explanations.

that a certain answer to a question Q is correct, even though that answer is inconsonant with other views he holds. He may accept that answer, on the basis of r, *despite* his other views. If so, there is a sense, and I think an important one, in which although he accepts an answer to Q as correct, which indeed it is, he does not understand q, simply because the answer is inconsonant with other views he holds. What Bromberger says seems to preclude this sense of not understanding.

There is a further problem. Bromberger does not attempt to explicate "A understands q," and, from what he says, it is not clear that we can simply take the negation of his explication of "A does not understand q." Consider the question "Is Concord the capital of New Hampshire?" Suppose I know the correct answer. If so, then at least one of the answers I can conceive is an answer I can accept, and at least one of the answers I can conceive is the right answer. So neither of Bromberger's conditions (1) nor (2) holds for me. Can it therefore be concluded that I understand whether Concord is the capital of New Hampshire? What Bromberger says appears to disallow speaking of understanding in such a case, since he claims that we cannot say that someone does not understand this.

In what follows, then, I want to propose an alternative account of understanding which, at some points, will be similar to Bromberger's, at others not. I shall treat "understanding" rather than "not understanding" as the more fundamental notion and propose conditions that are satisfied if A understands q (where q is the *oratio obliqua* form of a question Q). The conditions that follow are not independent of each other. But it will be useful to formulate them as I do so that later we may be able to consider ways in which A might not understand q.

1. Q is a sound question and A believes that it is. (By a sound question I mean one that admits of a correct answer that does not challenge the legitimacy of the question.) It is inappropriate to describe a certain physicist as understanding why atomic nuclei contain no neutral particles if in fact they do. There might be a theory which says that atomic nuclei contain no neutral particles and our physicist might understand this theory. But from this we cannot conclude that he understands why atomic nuclei contain no neutral particles. Moreover, if A does not believe that Q is a sound question, then he cannot be said to understand q. If an astronomer believes that the earth was not formed from the sun, then we cannot say that he understands how the earth was formed from the sun.

2. A has at least some acquaintance with the item(s) mentioned in, or presupposed by, Q. If A understands the manner in which electrons revolve about the atomic nucleus, then he is aware that there are

electrons, that there is an atomic nucleus, and that the former do revolve about the latter. If q is what is or was said or meant by something or someone, e.g., a sentence, statement, word, concept, then A has read or heard or somehow come across the sentence, statement, word, concept. If I am not aware that anyone spoke, I cannot be described as understanding what someone said when he spoke. This condition is not implied by the first one, since, e.g., a layman who has never studied physics may believe on the authority of the physicist that a certain question Q in physics is sound, even though he is not acquainted with the item(s) mentioned in, or presupposed by, Q.

3. A knows a correct answer to Q, and one that is consonant with his other views; he believes that this answer is correct and is consonant with his other views; and he knows what the answer means. If A does not know a correct answer to the question "How was the solar system formed?" we cannot say that he understands how the solar system was formed. Even if he knows a correct answer, but that answer is not consonant with his other views, we cannot say that he understands this. Recall the earlier example cited as a counter-example to Bromberger, in which someone knows how a prisoner escaped (by sliding underneath the door) but does not understand it, since it is not consonant with his other views. If A knows an answer to "How did the prisoner escape?" which happens to be correct though A believes that it is not, or if A believes that such an answer is not consonant with his other views, then either of these facts would also count against the claim that A understands how the prisoner escaped. Finally, A must not merely be able to *say* what answer is correct, he must know the meaning of what he is saying.

Here and earlier I have spoken of consonance. This can be construed in terms of probability in the light of one's views. The more improbable p is, given A's views, the less consonant p is with those views. Since consonance admits of degrees, what we should say is: "If A knows a correct answer to Q, but that answer is not consonant with his (other) views, to the extent that it is not, or to the extent that it is not according to A, he does not (fully) understand q."

4. To express the fourth condition requires a distinction between two types of questions. Some (sound) questions are of a type such that if someone were unable to conceive of an answer to a question of this type which is not inconsonant with his views,[4] the natural

---

[4] This includes the case in which he is unable to conceive of any answer at all as well as the case in which he is able to conceive of answers but all are inconsonant with his views.

inference is that he does not believe the question is (completely) sound, or he does not (fully) know what the question means, or he can conceive of several answers to the question, knows that one of them (but not which) is correct, and yet none is consonant with his views. Examples of such questions are: "How many people are there in this room?", "How high is Mont Blanc?", "Is Concord the capital of New Hampshire?" Someone might not know correct answers to such questions. But if he cannot even conceive of any possible answer to them which is not inconsonant with his views, then the natural inference is that one of the previous conditions obtains. If someone cannot conceive of an answer to the question "Is Concord the capital of New Hampshire?" that is not inconsonant with his views, either he believes the question unsound (e.g., he thinks New Hampshire is a city); or he does not know what it means; or he can conceive of several answers, knows that one of them is correct, yet none is consonant with (all) his views (e.g., he believes that atlases provide reliable information about state capitals, yet has read in one that Concord is the capital of New Hampshire and in another that Manchester is).

If Q is a question of this type then, one might be tempted to conclude, we would not say that A understands q, even if he knows an answer to Q that is correct and consonant with his views (cf. Bromberger). For example, "A knows how high Mont Blanc is" sounds perfectly proper, "A understands how high Mont Blanc is" sounds less so. But such a conclusion is too hasty. Someone who sees A about to climb Mont Blanc with no equipment might say "A does not understand how high Mont Blanc is." We might reply, "He understands how high it is," and by doing so either rebut the suggestion that he cannot conceive that it might be more than 15,000 feet high, or claim that he recognizes the significance or implications of the fact that it is so high, or both. In general, if Q is a question of the present type, then we could say that A understands q only if either (1) we are rebutting the suggestion that A does not know an answer to Q that is correct and consonant with his views and cannot conceive of such an answer (either because he cannot conceive of the answer which is correct, or, if he can, because that answer is not consonant with his views), or (2) if we are making the claim that A recognizes the significance or implications of the fact that such and such an answer to Q is correct.

There is another type of question, however. This type of question is such that from the mere fact that someone is unable to conceive of an answer that is not inconsonant with his views, we are not entitled to draw the inference that he does not believe the question sound, or does not know what it means, or can conceive of several answers, knows that one (but not which) is correct, but none is consonant with his views. If someone cannot conceive of an answer to the question "How was the solar system formed?" that is not inconsonant with his views, we are not automatically entitled to draw this conclusion. For example, such a person may believe the question is sound, may know what it means, yet may not be able to generate answers to it, or may not be able to generate answers and know that among these is a correct one. If Q is such a question, it is not at all odd to say that A understands q. And if we do say this, we are not necessarily rebutting the suggestion or making the claim described in the previous paragraph.

According to the present condition, then, if it is correct to say that A understands q, then either Q is a question of the second type, or, if it is a question of the first type then "A understands q" is being used either to rebut a suggestion that A does not know and cannot conceive of an answer to Q that is correct and consonant with his views or to say that A recognizes the significance or implications of the fact that such and such an answer to Q is correct.

I have suggested four conditions typically satisfied if A understands q. What can we say about *not* understanding q? If any of the conditions are not satisfied, then it is not the case that A understands q. Sometimes this is sufficient for saying that A does not understand q. Thus, if Q is unsound, we may conclude that A does not understand q. For example, if a physicist claims to understand why atomic nuclei contain no neutral particles, we might reply that he does not understand q at all. More typically, however, our requirements for not understanding are stricter. When we say "A does not understand q" we imply or presuppose that Q is a sound question, that A has some acquaintance with the item(s) mentioned in Q, and that A is not in the position of being able to conceive of several answers to Q that are consonant with his views, knowing that one is correct but not which. If we claim that A does not understand what B said, the implication is that B did say something, that A heard or read what B said,[5] and that A is not in the position of being able to conceive of several things B might have said, knowing that he said one of them but not which (recall the example of not understanding the metaphysical sentence in Bradley).

I think that we use "A does not understand q" in both ways described above; i.e., (a) where not satisfying some condition for

[5] See example (ii) in Sect. III.

understanding q is sufficient for not understanding q, or (b) where this is sufficient provided that Q is a sound question, A has some acquaintance with the item(s) mentioned in Q, and A is not in the position of being able to conceive of several answers to Q that are consonant with his views, knowing that one is correct but not which. In the former sense we can say that Lincoln did not understand why President Kennedy was assassinated; in the latter sense we cannot. It might be claimed that (b) represents a more strict and proper use of "A does not understand q," whereas (a) should be read "It is not the case that A understands q." But perhaps this is overly rigid.

If I am right, then, even in sense (b) there are several ways in which A might not understand q, i.e., several circumstances in which it would be proper to say that A does not understand q. For example, he might not believe Q to be a sound question (a violation of the condition of item 1 above). But most of the circumstances center around condition 3. So that we may employ sense (b), suppose we assume that Q is a sound question, that A has some acquaintance with the item(s) mentioned in Q, and that he is not in the position of being able to conceive of several answers to Q that are consonant with his views, knowing that one is correct but not which. Suppose further that Q is a question of the second type described in condition 4. Given these assumptions, A might be in any one of the following circumstances in virtue of which we could conclude that he does not understand q: he might be unable to conceive of any answer to Q; he might be able to conceive of several answers to Q but none of which is correct; he might be able to conceive of an answer to Q which is correct but which he believes to be incorrect; he might be unable to conceive of any answer to Q which is consonant with his views; and so forth. As already noted, understanding, and therefore not understanding, admits of degrees. The extent to which someone does not understand q can depend upon the extent to which a correct answer to Q is not consonant with his views. It can also depend upon the extent of his inability to conceive of such an answer. The more difficult A finds it to conceive of any answer, or of an answer that is consonant with his views, the greater the lack of understanding we may attribute to him.

Let me return now to explanation. We began with this schema:

(I) A would attempt to explain q to those in situation S by citing E, or by citing a number of things of which E is the most central.

Invoking the concept of understanding, I said that this could be expressed as follows:

(II) A would attempt to render q understandable to those in situation S by citing E, or a number of things of which E is the most central, as providing what A believes is or might be a correct answer to Q.

I then defined the concept of explanation by reference to (I) as follows:

(III) Given C, the context of classification, E is classifiable as being or providing an explanation of q, within situation S, if and only if there is (was, might have been) some (type of) A who would attempt to explain q to those in S by citing E or by citing a number of things of which E is the most central. Whether we choose "is," "was," or "might have been," and what type of A is allowed (e.g., leading scientist A) depends on C.

Let me now tie (I), (II), and (III) to my remarks about understanding.

Suppose A is attempting to render q understandable to those in S, where q is the *oratio obliqua* form of a question Q of the second type described in condition 4. Appealing to my account of understanding, this means that A is attempting to provide a correct answer to Q in order to enable those in S to satisfy the conditions for understanding described above. It does not necessarily follow that such an attempt will produce understanding of q by those in S. The answer A provides may not be correct, or may not be consonant with the views of those in S. Even if it is, once given such an answer, those in S may not satisfy other conditions for understanding q. So, to produce understanding, A may need to do more than provide a correct answer to Q. He may need to show that Q is a sound question. He may need to make those in S acquainted with items mentioned in, or presupposed by, Q. He may need to modify or augment the views of those in S in such a way that the answer becomes consonant with their views. If A does provide this additional information we can say, if we wish, that it is part of A's answer to Q. Let me call an answer to Q which does not contain any of this additional information a *narrow* answer to Q, and an answer to Q which does, a *broad* answer to Q. To provide a narrow answer to Q is to "just answer the question," without trying to show that it is sound, saying what it means, trying to show why the answer is plausible, and so forth. To provide a broad answer is to do some of these other things as well. In attempting to explain q to those in situation S, where Q is a question of the second type described in condition 4, A may cite E as providing either a broad or a narrow answer to Q. Depending on the S, either a broad or narrow answer to Q may render q understandable to those in S. It does not follow, of course, that it will.

Suppose now that Q is a question of the first type described in condition 4. Then we would say that those in S understand q only if we were either rebutting the suggestion that they do not know and cannot conceive of a correct and plausible answer to Q, or claiming that they recognize the significance or implications of the answer that is correct. So if A is attempting to explain q to those in S, i.e., if he is attempting to bring them to understand q by citing E, the implication may be that not only do those in S not know an answer to Q which is both correct and consonant with their views but that they cannot conceive of such an answer. This may be because they believe Q to be unsound, because they do not know what it means, because, although they can conceive of several answers to Q and they know that one of them is correct (but not which), none is consonant with their views, or because the correct answer to Q is not consonant with their views. Or else the implication may be that those in S do not recognize the significance of the fact that such and such an answer is correct. Accordingly, if Q is a question of the first type described in condition 4, and we want to say that E is or provides an explanation of q, or that A would attempt to explain q by citing E, then E contains not only (what A takes to be) a correct and plausible answer to Q, but also information provided in the light of the fact that those in S are (assumed to be) in one or more of the circumstances described above. For example, we might say that within S, A attempted to explain how high Mont Blanc is by citing E, and we might speak of E as being or providing an explanation of this, within S. If we do say this, then E will contain more than the information "15,781 feet." E will contain at least some information purporting to demonstrate the soundness of the question, or to indicate something about Mont Blanc in addition to its height, or to indicate the plausibility of the answer "15,781 feet," or to indicate the significance of this answer for the prospective climber. The proposition "Mont Blanc is 15,781 feet high" by itself is not, and does not provide, an explanation of how high Mont Blanc is. If, e.g., any answer to the question "How high is Mont Blanc?" that those in S can conceive is an answer that is not plausible, given their views, then we might explain to them how high Mont Blanc is by telling them that it is 15,781 feet and showing them why this figure is plausible (which, of course, will involve changing their views). In short where Q is a question of the first type described in condition 4, E will provide a broad rather than a narrow answer to Q.

We can bring all this together by writing down the following schema as an analysis of "explanation":

(IV) Given C, the context of classification, E is classifiable as being or providing an explanation of q, within situation S, if and only if there is (was, might have been) some (type of) A such that, within S, A would cite E as providing what he believes is or might be a correct answer to Q, where Q is a question of the second type described in condition 4, where E provides either a narrow or a broad answer to Q, and where A would provide such an answer in order to enable those in S to satisfy the previously cited conditions for understanding; if Q is a question of the first type, then E will provide a broad rather than a narrow answer to Q. (Whether we choose "is," "was," or "might have been," and what type of A is allowed, depends on C.)

IV. Evaluating Explanations

What does one evaluate when one evaluates an explanation? The E's I have spoken of as being or providing explanations can be such things as events, facts, phenomena, states of affairs, and propositions. These are items that A can cite in explaining q. But when I speak of evaluating an explanation I do not mean evaluating an event, fact, phenomenon, state of affairs, or proposition. What I mean is determining whether by citing one of these items A would be providing a good (satisfactory, illuminating, etc.) explanation of q. Following the procedure I have been adopting, this means evaluating A's answer to a question Q in which E is cited in order to render q understandable to those in S. To begin with, I shall note several criteria that are important for evaluating an explanation from the point of view of those in the situation S in which the explanation is offered. Afterwards I shall discuss the possibility of making evaluations independently of any S.

Persons in S either do not understand q, or can be treated as if they do not. What is not understood is, or can be expressed as, the *oratio obliqua* form of a question. However, if a person does not, or can be treated as if he does not, understand q, often there will be several questions he might raise in addition to Q.[6] If he does not understand why the spectrum of hydrogen contains discrete lines, he might ask how spectra are produced, what the structure of the hydrogen atom is, what general physical principles are involved, and so forth. With this in mind, I want to mention several considerations for deciding whether, by citing E, A is providing an explanation of q that is a good or satisfactory one within S.

(R_1) *Relevance*. By citing E, does A provide an answer to Q and to other questions that are or might be raised by those in S who do not, or can be treated as if they do not, understand q? Is A attempting

[6] This will always be true if Q is a question of the first type noted in condition 4 of the previous section.

to render q understandable, and to do so in respects that are or might be of concern to those in S?

(R_2) *Correctness.* By citing E, does A provide a correct answer to Q as well as to other questions that are or might be raised by those in S who do not understand q? How correct the answers need to be depends on standards appropriate for the situation: in some cases approximations suffice, in others not. Related criteria are plausibility, evidential support, and simplicity. (By citing E, does A provide plausible, evidentially well-supported, and simple answers to questions that are or might be raised in S?) Each of these is relevant in considering whether an explanation is correct, and each is used for purposes of evaluation. Of the three, perhaps plausibility is the broadest, being a function partly of evidential support, partly of simplicity, and partly of whether and to what extent the explanation is consonant with other acceptable ones.

(R_3) *Depth.* By citing E, at what level does A provide an answer to Q and to other questions that are or might be raised by those in S who do not understand q? Even though A does answer such questions, and his answers are correct (as far as they go), we may say that his explanation is shallow or superficial if we think that more fundamental considerations should have been adduced. Thus, although one can explain why gases at constant pressure expand by citing a temperature increase, a deeper answer would appeal to the molecular constitution of gases.

(R_4) *Completeness.* By citing E, how completely, at a given level, does A answer Q and other questions that are or might be raised by those in S who do not understand q? Even though A may attempt to answer questions about volume at the microscopic level, instead of the macroscopic (thermodynamic) level, his answers at the more fundamental level may be quite sketchy.

(R_5) *Unification.* To what extent does the citing of E provide unification for the items concerning which questions are or might be raised by those in S who do not understand q? An explanation may be given high marks in virtue of the fact that a few ideas help render understandable a wide range of items, relationships among which would otherwise be unknown or not considered.

(R_6) *Manner of Presentation.* When A presents his explanation does he do so in a sufficiently clear, simple, and organized manner so as to be likely to render q understandable to persons in S?

Here, then, are a number of relevant considerations for evaluating explanations from the point of view of those in S. Many terms of evaluation can be tied to one or more of them. We may speak of an explanation as a powerful one, on the basis of the unification it provides and in addition its depth. If we say that an explanation is illuminating, we may have in mind several criteria, especially relevance, depth, unification, and manner of presentation. If we say that it is reasonable, we will be employing one or more of the criteria mentioned under correctness. This is not to deny that certain terms, e.g., "original," "imaginative," and "influential," not tied to any of the criteria above can be used to evaluate explanations. I have not exhausted the dimensions in which evaluations are possible.

How are (R_1)–(R_6) to be used? When we evaluate an explanation we may do so only in certain respects. We may say that the Ptolemaic explanation of the motions of the planets is a good one with respect to completeness and unification, although it is not correct. Now within S an explanation satisfies (R_1) if and only if it provides an answer to Q and to other questions that are or might be raised by those in S who do not understand q; it satisfies (R_2) if and only if it supplies correct answers; and so on. What counts as relevant, how much accuracy, depth, completeness, or unification is appropriate, and what type of presentation is in order, depends on, and varies with, the knowledge and concerns of those in S. We evaluate E with respect to (R_i) by considering what standard of (R_i) is appropriate in S and then determining how E measures up to that standard. So we can say:

(A) Given the knowledge and concerns of those in S, E is or provides an explanation of q which is a good (satisfactory) one, within S, with respect to (R_i) (relevance, correctness, etc.) if and only if E is classifiable as an explanation of q within S and E satisfies (R_i) to an extent appropriate in S.

Sometimes we are willing to evaluate an explanation not simply with respect to individual criteria, but "on the whole"; i.e., taking all the criteria into account. One thing we can say is this:

(B) Where E is classifiable as an explanation of q, within S, the more criteria appropriate to S which E satisfies, and the greater the extent to which it satisfies them to the level appropriate to S, the better is E as an explanation of q within S.

If, however, given S, we want a schema for an "absolute" evaluation of E on the whole, the matter becomes much less precise. Possibly we attach varying degrees of importance to the criteria cited, but how we do this, and combine such information with that indicating the extent to which E satisfies each criterion, is not at all clear. So, quite vaguely, supposing now that (R_1)–(R_6) represent a sufficient number of criteria to evaluate E on the whole, we can say this:

(C) Where E is classifiable as an explanation of q, within S, E is, on the whole, a good or satisfactory explanation of q, within S, if and only if, on the whole, E satisfies the (R_i) appropriate to S.

An ideal explanation with respect to (R_1)–(R_6), within S, might be defined thus:

(D) Where E is classifiable as an explanation of q, within S, E is an ideal explanation of q, within S, if and only if E satisfies each of the (R_1)–(R_6) appropriate in S to the fullest extent appropriate in S.

I have spoken about evaluating an explanation given the knowledge and concerns of those in S. Can we evaluate an explanation independently of any S? It might be argued that we can, at least to a large extent. True, it might be said, to evaluate E one must consider the question Q that E purports to be answering. And this question itself indicates at least something about the knowledge and concerns of those in a situation S. The question partially defines such a situation. But we can determine whether E provides an answer to Q that is relevant, correct, etc., without very fully specifying S. We can judge that one explanation is deeper than another by specifying a question it is attempting to answer, without indicating anything further about any S.[7]

The problem is that given just Q we cannot always determine the level of (R_i) that is appropriate, and this is necessary for evaluating E. Given just Q we may be able to say that E_1 is a deeper explanation than E_2. But can we conclude that it is better? Can we give E_1 higher marks, even if E_2 and E_2 are equal in other respects? The answer depends upon what level of depth is appropriate in S, which in turn depends on the knowledge and concerns of those in S, something not necessarily indicated by Q. An explanation of a blow-out in an automobile tire that makes reference to kinetic theory is deeper than one appealing only to the expansion of the air as a result of temperature increase. But it is not necessarily better. Here we do have to

[7] It might be objected that this is not really consistent with the policy of relativizing the definition of explanation to an S. So two moves might be made. When we consider the extent to which E satisfies (R_i) we might assume that E is to count as an explanation of q within some S, which is not further specified. Or we might speak of E as a potential explanation of q and judge that one potential explanation is deeper than another, independently of any S.

consider the knowledge and concerns of those in S, information not fully indicated by the question "Why did the tire blow-out?" The deeper explanation is not always the better one. Similar remarks apply to completeness and unification. With relevance and correctness matters are somewhat different, for in all situations these are important in evaluating an explanation on the whole. Still, what counts as relevant depends upon S, and what degree of correctness is appropriate varies with the situation. Finally, what manner of presentation is appropriate will also depend to some extent on S. In certain situations a rigorous axiomatic presentation may be in order, in others a much less formal, and more "intuitive" one.[8]

It might be objected that explanations in science, at least, can be evaluated knowing just the questions they are attempting to answer. Since science aims at correctness, depth, completeness, and unification, the more accurate, deep, complete, and unifying explanation is always the better one, scientifically speaking.

In science, to be sure, the situations presupposed when explanations are given are considerably standardized. A text in kinetic theory does not offer a different explanation of gaseous phenomena for each reader with a different interest and background. Rather it presupposes overall uniformity in its audience with respect to knowledge and interests. Nevertheless, different texts will develop kinetic theory at different levels, with varying degrees of accuracy, depth, completeness, and unification; even the same text may do so in different chapters. Most important, it is simply untrue that we can evaluate an explanation of q, even in science, just by considering the explanation and the q. The appropriate level of correctness, depth, and completeness must be considered as well. At least three factors are responsible for this. In science there is often a quest for simplicity in explanation, for familiarity, and variety. Each of these desiderata can affect the amount of accuracy, depth, completeness, and unification that is appropriate, which in turn affects our evaluation of the explanation. For example, in kinetic theory, why monatomic gases have the specific heats they do can be explained by ignoring rotation of molecules, collisions between molecules, and quantum mechanical effects, although all of these exist. The explanation proceeds by treating gas molecules as structureless mass points with translational energy only. Admittedly, it provides an answer that describes gases in a manner that is not very accurate, complete, or deep. Still, these additional factors have little effect on the specific heats of monatomic gases.

[8] See my *Concepts of Science* (Baltimore, 1968), chap. 4.

In many scientific situations what is wanted is a simple explanation that takes into account the principal factors only. This is not to deny that in some scientific situations greater accuracy, depth, or completeness may be required. It is also typical in science to approach an issue from several points of view. In explaining transport phenomena, i.e., viscosity, heat conduction, and diffusion in gases, the physicist may use mean-free-path methods based on the idea of a billiard ball model in which molecules travel certain distances (free paths) between collisions. He may use momentum transfer methods in which the net transfer of momentum during molecular collisions is considered. Or he may use methods based on the distribution of molecular velocities. Each of these explanations can provide important insights concerning transport phenomena, and texts on kinetic theory frequently include all of them. The first two have the advantages of simplicity and familiarity, the latter is more rigorous. Which explanation to rate as the best depends on the knowledge and concerns of those in the situation in which it is to be invoked. In general, the fact that E_1 is more accurate, deeper, more complete, or more unifying than E_2 does not, without at least implicit appeal to a type of situation S, permit us to say that it is a better explanation. It permits us to say only that in, or relative to, certain situations it is a better explanation.[9]

The Johns Hopkins University

[9] I am indebted to the National Science Foundation for support of research.

II
Theoretical Terms and Inductive Inference

KEITH LEHRER

ONE purpose of a scientific theory is to systematically relate terms that describe observable states of the world. Such terms, observation terms, are distinguished from those theoretical terms that describe unobservable theoretical states. The systematizing of observation terms is achieved by the use of theoretical terms. It is agreed that theoretical terms enable us to achieve this sort of systematization in an economical way and also that they have great heuristic value in science. However, it is natural to think that theoretical terms are not only used economically and fruitfully in effecting such a systematization, but are logically indispensable for this purpose. My objective is to explain how this can be so.[1]

First I shall have to distinguish two kinds of systematization: inductive and deductive. To this end, let us define an empirical statement as one that contains only observation terms in addition to logical and mathematical terms and that is contingent, neither L-true nor L-false. Then we may say that a system achieves *deductive* systematization if the system in conjunction with some empirical premiss enables us to deduce some empirical statement as a conclusion that would not have been deducible from the empirical premiss alone. Similarly, a system achieves *inductive* systematization if the system in conjunction with some empirical premiss enables us to induce some empirical conclusion that would not have been inducible from the empirical premiss alone. Finally, I shall say theoretical terms are logically indispensable to a systematization if and only if they achieve systematization and there is no alternative system achieving the same systematization without the use of theoretical terms.

---

[1] An earlier version of this paper was presented at the 3rd International Congress for Logic, Methodology, and Philosophy of Science, and was written while the author was an NSF Senior Postdoctoral Fellow at The University of Edinburgh. The paper has been altered in response to the suggestions and criticisms of H. E. Kyburg, Rolf Eberle, Carl Hempel, Mary Hesse, and an anonymous referee of this journal. I do not suggest that they would subscribe to the result.

Carl G. Hempel in an article entitled, "The Theoretician's Dilemma," has contended that theoretical terms are not logically indispensable, however useful they might be, for the purpose of producing deductive systematization.[2] Nevertheless, Hempel argues that theoretical terms may be logically indispensable for inductive systematization.[3] This conclusion is one I wish to defend, though I shall argue that Hempel's support of it is inconclusive. My thesis is that if theoretical terms and observation terms are connected by the usual logical connectives, then, even if the system in which they are embedded yields systematization, the theoretical terms may still be logically dispensable. On the other hand, if the observation terms and theoretical terms are connected by a relation of probability, then the theoretical terms can prove logically indispensable to the systematization achieved.

I

To establish my conclusions, I shall consider three very simple systems containing theoretical terms which achieve some systematization. All the systems contain a theoretical vocabulary consisting of a primitive theoretical predicate 'T' and two primitive observation predicates "O_1" and "O_2." Let us first consider a system effecting deductive systematization and recall Hempel's argument.[4]

System S1
P1. $(x)(O_1 x \supset Tx)$
P2. $(x)(Tx \supset O_2 x)$.

We may think of S1 as part of a larger system containing purely theoretical statements as well as statements like P1 and P2 which relate theoretical and observation predicates. Hempel refers to the latter sort of statements as interpretative, and I shall consider P1 and P2 to be meaning postulates for the theoretical term 'T.' From this system we may deduce the empirical statement:

E1. $(x)(O_1 x \supset O_2 x)$.

Moreover, every empirical statement that may be deduced from S1 may be deduced from E1. Therefore, E1 effects precisely the same

[2] Carl G. Hempel, "The Theoretician's Dilemma" in *Minnesota Studies in the Philosophy of Science*, ed. by Herbert Feigl, Michael Scriven, and Grover Maxwell, vol. II (Minneapolis, University of Minnesota Press, 1958), pp. 37–99, esp. pp. 75–78.

[3] *Ibid.*, pp. 78–80 and Carl G. Hempel, "Implications of Carnap's Work for the Philosophy of Science" in *The Philosophy of Rudolph Carnap*, ed. by Paul Arthur Schilpp (La Salle, Illinois, Open Court, 1963), pp. 685–707, esp. pp. 700–701.

[4] Hempel, "Dilemma," *op. cit.*, pp. 78–80.

deductive systematization of observation terms as S1. The theoretician's dilemma arises from a generalization of this result due to William Craig to the effect that for any system S containing theoretical terms, like S1, there is a system S_0 containing no theoretical terms, like E1, such that S and S_0 effect precisely the same deductive systematization of observation terms, that is, they have the same empirical statements as deductive consequences.[5] Hence the dilemma. Either a system containing theoretical terms effects deductive systematization or it does not. If it does not, then obviously the theoretical terms are not indispensable for that purpose. But, even if the system does achieve deductive systematization, the theoretical terms are not logically indispensable. So, either way, theoretical terms are logically superfluous for this purpose.

II

To show how theoretical terms can be logically indispensable for inductive systematization, Hempel asks us to consider a system that does not achieve any deductive systematization. I shall examine a simpler system that generates the same issues.

System S2
P1. $(x)(O_1x \supset Tx)$
P2. $(x)(O_2x \supset Tx)$.

This system does not produce any deductive systematization; no empirical statement is deducible from it. However, it might be argued that the system does achieve some inductive systematization. Suppose that we observe "b is O_1," and from S2 deduce "b is T." From this conclusion we cannot deduce "b is O_2." However, it might seem reasonable to induce "b is O_2."[6] Moreover, even if one doubts that it would be reasonable to induce the conclusion "b is O_2" from the premiss "b is T," nevertheless, given P2, the statement "b is T" is at least positively relevant to the statement "b is O_2." By saying that the statement "b is T" is positively relevant to the statement "b is O_2," I mean that the conditional probability of the second on the basis of the first is higher than the antecedent probability of the second. The proof of positive relevance is as follows. Reading "$P(B, A)$" as "The probability of B on the condition that A" and "$P(A)$" as "The antecedent probability of A," it follows from the calculus of probability that $P(O_2b, Tb) = P(O_2b \ \& \ Tb)/P(Tb)$, and,

[5] Ibid.
[6] The examples presented by Hempel in "Dilemma," op. cit., pp. 78–79, and "Implications," op. cit., pp. 700–701, should be compared because the systems he formulates differ in logical form from S2.

with P2, that $P(O_2b \ \& \ Tb) = P(O_2b)$. Thus $P(O_2b, Tb) = P(O_2b)/P(Tb)$.[7] Therefore, assuming that $P(Tb)$ is less than 1, $P(O_2b, Tb)$ is greater than $P(O_2b)$, which means that the statement "b is T" is positively relevant to the statement "b is O_2."

If the statement "b is T" is thus positively relevant to the statement "b is O_2," it might be conceded that given S2 the statement "b is O_1" is positively relevant to the statement "b is O_2," because given S2 we can deduce "b is T" from the premiss "b is O_1." The latter would not have to be conceded. In this example, as in the ones Hempel gives, one could quite reasonably deny that the system establishes the positive relevance of any empirical statement to any other. There is no theorem of probability that would justify an inference from S2 and the fact that "b is T" is positively relevant to the statement "b is O_2" to the conclusion that the statement "b is O_1" is positively relevant to the statement "b is O_2."[8] However, even if that conclusion is conceded, the theoretician is still in the woods. The logical indispensability of theoretical terms for inductive systematization cannot be established on that basis.

To see why not, let us suppose we may formulate the positive relevance conceded in terms of a probability statement to the effect that the probability that x is O_1 on the condition that x is O_2 is at least n as follows:

E2. $P(O_2x, O_1x) = n$,

where the value of n is greater than the probability of m in the probability statement:

$P(O_2x) = m$.

Let us assume that the probability statement E2 establishes the same inductive relations between empirical statements as does S2. Given this assumption, we can easily obtain a dilemma pertaining to the inductive systematization corresponding to the dilemma pertaining to deductive systematization. For, E2 would yield precisely the same inductive systematization as would S2. Thus, if we allow the probability functor as part of our logical-mathematical machinery, then we can conclude that there is a system without theoretical terms which effects the same inductive systematization as does S2, namely, the system consisting of E2. Just as E1 gave us the same deductive systematization as S1, but without employing theoretical terms, so E2 gives us the same inductive systematization as S2, again without the

---

[7] All that I am assuming about the probability functor is that it satisfies the elementary axioms of any familiar calculus of probability.

[8] There are counterexamples to inferences of this general form.

use of theoretical terms. The theoretical terms in S2 are, therefore, no more logically indispensable for the inductive systematization they achieve than the theoretical terms in S1 are logically indispensable for the deductive systematization they achieve. From the standpoint of systematization the theoretical terms of both systems are logically otiose.

This result concerning the system S2 naturally raises the question as to whether it might be possible to generalize the result as it was possible to generalize the result concerning S1. If so, we could prove that for any system containing theoretical terms, like S2, which effects inductive systematization, there is another system, not containing theoretical terms, which effects the same systematization. Could this be proven, then we would be forced to conclude that theoretical terms are logically superfluous for the purpose of systematization, be it deductive or inductive. I shall now show that this general result cannot be established.

III

If we construe the connection between observation terms and theoretical terms as a relation of probability, then it is possible to construct a system in which the theoretical terms are logically indispensable for producing inductive systematization.[9] Consider the following system incorporating this proposal:

System S3
P1. $P(Tx, O_1x) = r$
P2. $P(O_2x, Tx \ \& \ O_1x) = s$.

I am here assuming the following:
$P(Tx) = p (p < r)$
$P(O_2x) = q (q < s)$.

Thus, the probability statements guarantee the positive relevance of the terms related by the functor. The system establishes the positive relevance of theoretical statements to observation statements and vice versa. In virtue of S3 the statement "b is O_1" is positively relevant to the statement "b is T," and the statement "b is T and O_1" is positively relevant to the statement "b is O_2."

However, the most important feature of S3 is a negative one.

[9] Michael Scriven, "Definitions, Explanations, and Theories" in *Minnesota Studies in the Philosophy of Science, op. cit.*, pp. 180 ff., discusses the role of probability in connecting theoretical and observation terms, but he seems ultimately to reject the significance of the distinction between these two kinds of terms. Thus his views may well be inconsistent with mine in spite of some superficial similarity.

From S3 it does not follow that the statement "b is O_1" is positively relevant to the statement "b is O_2." It is consistent with this system to assume that the $P(O_2x, O_1x)$ is less than r or s, and, consequently, that $P(O_2x, O_1x) = P(O_2x)$. From the calculus of probability it follows that $P(O_2x \,\&\, Tx, O_1x) = P(Tx, O_1x) \times P(O_2x, Tx \,\&\, O_1x)$. That $P(O_2x \,\&\, Tx, O_1x) = P(O_2x, O_1x) \times P(Tx, O_2x \,\&\, O_1x)$, also follows and hence $P(O_2x \,\&\, Tx, O_1x) = P(O_2x, O_1x)$ if $P(Tx, O_2x \,\&\, O_1x) = 1$. We can insure that this is the case by adding

P3. $(x)((O_1x \,\&\, O_2x) \supset Tx)$

to S3. In this case, $P(O_2x, O_1x)$ is less than either r or s (assuming both are less than 1), and we may therefore consistently assume that $P(O_2x, O_1x) = P(O_2x)$. It would then follow that $P(O_1x, O_2x) = P(O_1x)$ because $P(O_1x, O_2x) = P(O_1x) \times P(O_2x, O_1x)/P(O_2x)$. Therefore, it is consistent with S3 to assume that the statement "b is O_1" is *not* positively relevant to the statement "b is O_2" and vice versa. Thus, if S3 achieves any inductive systematization between "O_1" and "O_2," then, on the assumptions we have made, the same systematization is not achieved by a system of probability statements asserting the positive relevance of one of those terms to the other.

But does S3 achieve any inductive systematization? We have said that a system achieves inductive systematization if the system in conjunction with some empirical premiss enables us to induce an empirical conclusion that would not have been inducible from the empirical premiss alone. Therefore, to prove that S3 does achieve some inductive systematization among empirical statements, we must prove that from some empirical premiss in conjunction with S3 we can induce an empirical conclusion that we cannot induce from that premiss alone. To prove this we must first formulate some inductive rule that would enable us to induce conclusions from premisses.

IV

The problem of formulating a satisfactory inductive rule is currently a matter of investigation and controversy.[10] However, any

---

[10] Articles opposing the idea that there should be acceptance rules are: Rudolph Carnap, "The Aim of Inductive Logic" in *Logic, Methodology, and Philosophy of Science*, ed. by Ernest Nagel, Patrick Suppes, and Alfred Tarski (Stanford, Stanford University Press, 1962), pp. 303–318; and R. C. Jeffrey, "Valuation and Acceptance of Scientific Hypothesis," *Philosophy of Science*, vol. 23 (1956), pp. 237–246. Acceptance rules are proposed in R. M. Chisholm, *Perceiving: A Philosophical Study* (Ithaca, Cornell University Press, 1957); C. G. Hempel, "Deductive-nomological versus Statistical Explanation" in *Minnesota Studies in the Philosophy of Science*, ed. by Herbert Feigl and Grover Maxwell, vol. III (Minneapolis, University of Minnesota Press, 1962), pp. 98–169; H. E. Kyburg, "Probability, Rationality and a Rule of Detachment" in *Proceedings of the 1964*

reasonable rule of this sort must at least preserve consistency in the sense that it must not allow us to induce inconsistent conclusions from consistent premisses. Thus, though an inductive rule, unlike a deductive rule, might permit the inference of false conclusions from true premisses, it must not permit the inference of inconsistent conclusions from true premisses. Many rules that it would seem natural to accept lead to such inconsistencies. For example, it is natural to suppose that there is some number m/n less than 1 (where m and n are positive integers) such that if e is our total evidence, or all the evidence we have relevant to h, and $P(h, e)$ is at least m/n, then we may induce h from e. But any rule of this form will lead to the result that inconsistent conclusions are inducible from consistent premisses. To see that this is so, suppose that there is a set of hypotheses $h_1, h_2, \ldots, h_n$ that are equiprobable with respect to evidence e, and are such that it follows from e that exactly one of the hypotheses is true. In this case, $P(-h_1, e)$ is at least m/n, and the probability of the denial of each of the other hypotheses is the same with respect to evidence e. Thus, given a rule of the kind in question, we could induce each of the negative hypotheses $-h_1, -h_2, \ldots, -h_n$ from e. But we could induce also $-(-h_1 \& -h_2 \& \ldots \& -h_n)$ from e, because the probability of the denial of this conjunction is 1. Therefore, we could induce an inconsistent set of conclusions from a consistent premiss. Hence, rules of this sort, which specify some degree of probability as sufficient for inducing a conclusion from a premiss, must all be rejected as unsatisfactory.

One way to avoid such unsatisfactory results is to adopt a rule insuring that the conjunction of the set of hypotheses induced from a premiss is at least as probable as one of the hypotheses induced and thus that the conjunction rather than its denial may be induced from the premiss. I shall propose such a rule for application to finite languages. (The elaboration of the rule to make it applicable to infinite languages would involve complexities not relevant to the present issue.) In formulating the rule, I shall want to distinguish between those hypotheses that may be directly induced from a premiss and those that may be indirectly induced from that premiss. I shall use the notation "$D(h, e)$" to mean "h may be directly induced from e," and the rule for direct induction is as follows:

International Congress for Logic, Methodology, and Philosophy of Science, ed. by Y. Bar-Hillel (Amsterdam, North-Holland, 1965), pp. 303–310; Jaakko Hintikka and Risto Hilpinen, "Knowledge, Acceptance, and Inductive Logic" in *Aspects of Inductive Logic*, ed. by Patrick Suppes and Jaakko Hintikka (Amsterdam, North-Holland, 1966); and Isaac Levi, *Gambling with Truth* (New York, Alfred Knopf, 1967).

(RDI) If, for any h that is not a deductive consequence of the conjunction k & e, $P(k, e)$ exceeds $P(h, e)$, then $D(k, e)$ provided that e is consistent with k.

In finite languages $P(h, e)$ is 1 if and only if h is a deductive consequence of e. Thus, (RDI) only requires that a directly induced hypothesis be more probable than those hypotheses having a probability less than 1 with respect to the conjunction of e and the induced hypothesis.

The set of statements h such that $D(h, e)$ is established by (RDI) is logically consistent and the probability of the conjunction of such hypotheses is equal to the probability of one of the conjuncts. Both these features follow from that fact that for any two hypotheses h and k such that $D(h, e)$ and $D(k, e)$ is established by (RDI), either h is deducible from the conjunction of k and e, or k is deducible from the conjunction of h and e. Moreover, though the rule is quite restrictive, a less restrictive rule would allow the possibility that there should be two hypotheses h and k such that $D(h, e)$ and $D(k, e)$, but such that $P(h$ & $k, e)$ is less than $P(h, e)$ and less than $P(k, e)$. This can be seen from the fact that $P(h$ & $k, e) = P(h, e) \times P(k, h$ & $e) = P(k, e) \times P(h, k$ & $e)$. Thus, either $P(k, h$ & $e) = 1$, in which case k is deducible from h & e, or $P(h, k$ & $e) = 1$, in which case h is deducible from k and e, or $P(h$ & $k, e)$ is less than $P(h, e)$ and also less than $P(k, e)$.

Moreover, we need not restrict the conclusions induced from a premiss to those directly induced. Having seen that (RDI) will not lead us to inconsistent conclusions from consistent premisses, let us adopt the following rule of induction:

(RI) A hypothesis h may be induced from evidence e if h belongs to A_e.

Let
$$A_e = A_1 \cup A_2 \cup \cdots \cup A_n$$
where
$A_1 =$ the set of hypotheses h such that $D(h, e)$ by (RDI),
and, letting a_i be a conjunction logically equivalent to A_i,
$A_2 =$ the set of hypotheses h such that $D(h, e$ & $a_1)$ by (RDI)
$A_3 =$ the set of hypotheses h such that $D(h, e$ & $a_2)$ by (RDI)
.
.
.
$A_n =$ the set of hypotheses h such that $D(h, e$ & $a_{n-1})$ by (RDI).[11]

[11] This formulation of (RI) was suggested to me by Rolf Eberle.

Thus A_e is the set of hypotheses which includes all those hypotheses that we may directly induce from e by (RDI) and all those hypotheses we may directly induce from the set of former hypotheses in conjunction with e and so forth. These hypotheses taken together constitute the set A_e of consequences we may induce directly or indirectly from e. The conjunction of all these hypotheses is also consistent and, thus, by using (RI) we may avoid inducing inconsistent conclusions from consistent premises.

V

Adopting (RI) as our rule of induction, let us reconsider S3 where

P1. $P(Tx, O_1x) = r$
P2. $P(O_2x, Tx \& O_1x) = s$

and where we also suppose that

P3. $(x)((O_1x \& O_2x) \supset Tx)$

so that $P(O_2x, O_1x) = P(Tx \& O_2x, O_1x) = r \times s$. Can we prove that from S3 in conjunction with some empirical premiss we can induce an empirical conclusion that we cannot induce from the empirical premiss alone? To do so we shall need to add some further assumptions.

Assume that r is high enough so that $P(Tb, O_1b)$ is greater than $P(h, O_1b)$ for any h that is not deducible from "$O_1b \& Tb$." In such a case we may directly induce "Tb" from "O_1b" by using (RDI). Secondly, assume that s is sufficiently high so that $P(O_2b, Tb \& O_1b)$ is greater than $P(h, Tb \& O_1b)$ for any h that is not deducible from "$O_2b \& Tb \& O_1b$." In this case we may directly induce "O_2b" from "$Tb \& O_1b$." Therefore, given S3 are a set of interpretative sentences or meaning postulates for T, we may induce "O_2b" from "O_1b" by our rule of induction. Since the assumptions concerning the values of r and s are consistent, we may conclude that S3 does achieve inductive systematization.

Having arrived at the foregoing result, we must now attempt to prove that the same systematization of empirical statements cannot be achieved without the use of theoretical terms. We have assumed that $P(O_2b, O_1b) = P(O_2b) = r \times s$. Given this assumption, it is consistent to further assume that there is some hypothesis h that is not deducible from "$O_2b \& O_1b$" such that $P(h, O_1b)$ is equal to or greater than $P(O_2b, O_1b)$, and, moreover, that for any set of hypotheses A_i induced from "O_1b," there is a hypothesis h that is not deducible from "$O_2b \& O_1b \& a_i$" such that $P(h, O_1b \& a_i)$ is equal to or greater than $P(O_2b, O_1b \& a_i)$. In short, it is perfectly consistent to assume that probabilities are such that we cannot induce "O_2b"

from "O_1b" on the basis of probability statements relating only observation terms, but can induce "O_2b" from "O_1b" on the basis of probability statements such as those of S3 which relate observation and theoretical terms.

Under the assumptions we have made, S3 achieves inductive systematization and the same systematization would not be produced by a probability statement relating the observation terms "O_1" and "O_2." To prove that under these assumptions the theoretical term in S3 is logically indispensable to the systematization, we need only prove that it follows from these assumptions that there is no alternative system achieving the same inductive systematization without the use of theoretical terms. Since the systematization achieved depends essentially on certain probability relations, we may conclude that any alternative system, not containing theoretical terms, and achieving the same systematization, would have to consist of probability statements relating observation terms. However, given our assumptions, no system of probability statements relating observation terms can achieve the same inductive systematization as S3. Therefore, the theoretical terms in S3 are logically indispensable to the inductive systematization achieved by that system. This results because the theoretical term in S3 is related to observation terms by relations of probability rather than by the usual logical connectives.

VI

There are three objections to the foregoing argument that require comment. First, one might object to the inductive rule on which my argument has been based. I cannot here undertake to defend that rule against all alternatives, but it will suffice to add that other inductive rules, for example those which make induction depend on some specific degree of probability, could be substituted in the argument without loss of cogency. However, it must be conceded that any thesis concerning inductive systematization is likely to remain controversial until fundamental problems in inductive logic have been resolved.

Second, one might object that the theoretical term T in S3 is not indispensable to the inductive systematization achieved because that systematization could have been achieved by substituting some observation term "O_3" for T in the postulates of S3. The postulates would then be as follows:

System S4
$P(O_3x, O_1x) = r$
$P(O_2x, O_3x \& O_1x) = s$.

If we make the same assumptions regarding "O_3" that we made with respect to T, then we could argue that this alternative system, not containing theoretical terms, achieves the same systematization as S3. In particular, it allows us to induce "O_2b" from "O_1b" as did S3.[12]

The reply to this objection is that S4 does not achieve the same systematization as did S3 because S4 yields inductive systematization that S3 does not. Granted the assumptions indicated, we could induce "O_3b" from "O_1b" in conjunction with S4, but S3 would not warrant such an induction. Since S4 does not produce the *same* systematization as S3, but excessive systematization as well, the objection fails to provide a system without theoretical terms that achieves the same systematization as S3.

Third, it might be objected that a system consisting of the probability statement

E2. $P(O_2x, O_1x) = n$

where n is high enough to allow us to induce "O_2b" from "O_1b" by our rule of induction, is an alternative system containing no theoretical terms that achieves the same inductive systematization as S3. Such an objection would have to grant that E2 is not a consequence of S3 and that, given certain assumptions, S3 would be inconsistent with E2. Nevertheless, the objection would hold that E2 could achieve the same inductive systematization among empirical statements as S3.

My reply is twofold. This objection might well be rejected on the same basis as the preceding one. From the plausible assumption that $P(O_1x) = P(O_2x)$, it would follow from the calculus of probability that $P(O_1x, O_2x) = P(O_2x, O_1x)$, and, given E2, we could then induce "O_1b" from "O_2b" by our inductive rule. However, there is no reason to suppose that S3 would warrant this induction, and, indeed, it is consistent to assume that it would not. Again, the alternative system proposed would produce systematization not produced by S3, and therefore, it would not achieve the same systematization.

Second, if the objection merely asserts that there are conditions or assumptions under which E2 would achieve the same systematization as S3, I concede that this could be so. However, this does not refute what I have averred. I have argued that under certain conditions, ones that imply the falsity of E2, the theoretical term in S3 is indispensable for achieving inductive systematization among empirical statements. I concluded that under certain conditions theoretical statements are indispensable for achieving inductive systematization. There are no conditions under which theoretical terms are indis-

---

[12] This objection was raised by Professor Hempel in correspondence.

pensable for deductive systematization, and there are conditions under which theoretical terms are not indispensable for inductive systematization. But the fact that theoretical terms are sometimes superfluous is irrelevant to what I have proven, namely, that given perfectly consistent assumptions, theoretical terms are logically indispensable to achieving inductive systematization among empirical statements.

VII

Having achieved my purposes, I wish to conclude by maintaining that the system I constructed is a useful, if oversimplified, model of an essential ingredient in scientific theory. It is plausible to argue that inference from observable states of the world to theoretical states and vice versa is in many cases probabilistic. The parallel with quantum physics is obvious. Moreover, this model enables us to understand why theories are not rejected on the basis of a single counter instance. If the relations between theoretical terms and observation terms are implications, as they are in S1, then a single observed individual that is O_1 but not O_2 would refute the theory, and rationality would require that the theory be rejected. But, no matter how high the probabilities, if less than 1, it could be reasonable to retain system S3 even though we observed an individual that is O_1 but not O_2. On the other hand, when the number of such individuals increases, it would become increasingly unreasonable to retain S3, and, if we are tempted to save the theory by lowering the probabilities, we shall soon find that the inductive systematization of the system is lost. If, finally, it be asked how we are to decide how high to set the probabilities, the answer is simple. Set the probabilities as high as you reasonably can because by so doing you explain as much as you can and leave as little unexplained as you must.[13] Moreover, though systems so constructed are not immediately falsifiable by observation, they are not immune from experimental criticism, and by constructing a system that explains as much as possible, one thereby exposes one's system to experimental criticism. Thus, the construction I have offered to solve the theoretician's dilemma, is also intended to explicate the nature of rational scientific inquiry.

University of Rochester

---

[13] Cf. Wilfrid Sellars, "The Language of Theories" in *Science, Perception and Reality* (London, Routledge and Kegan Paul, 1963), pp. 106–126.

III
The Conventionality of Geometry

LAWRENCE SKLAR

I

THAT conventionalism is the last refuge of the *a priorist* is now a familiar theme in philosophy. It is curious, however, that the grounds philosophers have offered for claiming one or another science conventional are not of a single type at all. The over-worked litany that arithmetic is "merely conventional" or "solely a matter of language" has as its only motivation the empiricist maxim: if you can't possibly find anything in experience which would refute it, then it really doesn't say anything at all. No question of alternative arithmetics arises. Instead, it has proven a difficulty for the conventionalist view that there seem to be no plausible alternatives to ordinary, "old-fashioned" arithmetic, despite the alleged mere conventionality of this science.

In the case of geometry, on the other hand, it was just the discovery, in the mid-19th century, of alternative consistent three-dimensional geometries that first suggested the claim that the familiar Euclidean geometry was merely a conventional choice of theory. Here the motivation was different from that which prompted empiricists to label arithmetic "conventional." If there was, in fact, more than one consistent three-dimensional geometry, why had we believed, for such a very long time, not only that Euclidean geometry was the correct geometry to describe physical space, but that we had certain and indubitable grounds for making this assertion? Were great mathematicians and philosophers merely making a foolish mistake for all these centuries, a mistake only now to be corrected by modern thinking? Or was there some way in which one could, without denying the clear proof of the existence of alternative consistent three-dimensional geometries, still maintain that the familiar Euclidean geometry was the correct geometry for space, and could be known to be so *a priori*?

The motivation for the assertion that one's choice of geometry is conventional is now clear. If one can plausibly maintain that the choice of geometry is "merely a matter of convention," one can then show, not that Euclidean geometry is *the a priori* proper geometry for physical space, but that it is *a* proper geometry. That is, that one could correctly choose Euclidean geometry to describe physical space, and that one could do so with no fear of contradiction by experience, since one's choice was only a matter of convention. This now justifies the belief of generations of mathematicians and philosophers that one can know space to obey the laws of Euclidean geometry without reliance on observation or experimentation. The mathematicians were wrong, of course, if they also thought that this knowledge that space was Euclidean showed that any other geometry for space was inconceivable or inconsistent. For one could just as well, except for possibly unnecessarily complicating one's science, have chosen conventionally to "know" that space was non-Euclidean. But this latter error is not the error of thinking that one can know with certainty that space is Euclidean, for this belief is *not* in error. One can know, *a priori* and with certainty, that space is Euclidean, because one can conventionally *choose* Euclidean geometry as the geometry of space.

II

The line we have propounded above is, essentially, that taken by Poincaré in his famous essays on the epistemological status of geometry.[1] He added to it the assertion that, since Euclidean geometry was the simplest geometry, we would always choose it as the proper geometry of space. We need, he claimed, have no fear of any of its alternatives coming into favor. Poincaré, however, although the first to propound the claim that geometry was a matter of convention, insufficiently explicated the exact meaning of that claim. He did provide a famous two-dimensional "parable," though, which contains the essential arguments in a condensed and persuasive form.

Some time later, Reichenbach provided what is still the most thorough and lucid attempt to clarify the meaning of the assertion that geometry is conventional.[2] Although I will offer an explication which is critical of Reichenbach's position in several respects, the analysis of the doctrine of conventionality which I will propose is most deeply indebted to Reichenbach's work. For this reason I will postpone exposition of his doctrine until later. I will merely remark

---

[1] Henri Poincaré, "Non-Euclidean Geometries," "Space and Geometry," and "Experiment and Geometry" all in *Science and Hypothesis* (New York, Dover, 1952), Pt II.

[2] Hans Reichenbach, *The Philosophy of Space and Time* (New York, Dover, 1958), chap. I, sects. 1–8.

here that Reichenbach's theory, coming as it does after the publication by Einstein of his papers on general relativity, is much less committed to the eternal conventional choice of Euclidean geometry as the theory of physical space, than was the doctrine of Poincaré. For now the primary consideration appears to be not the simplicity of the geometric theory alone, but, rather, the simplicity of the total geometric-plus-physical theory of the world. And the simplicity of this total theory, according to Reichenbach, might be increased at the price of choosing a non-Euclidean geometry for the geometric portion of the total theory.

The path I intend to follow is this: first to examine, cursorily, some of the attempts made to analyze the meaning of the claim that geometry is conventional. For this has been a most peculiar chapter in the history of recent philosophy of science. We have had the spectacle of a whole procession of philosophers, each agreeing with his predecessors that geometry is conventional, but each proposing a radically different thesis as to just what the claim of conventionality for geometry amounts to. I do not wish to make this article primarily critical, nor to enter into a long discussion of the attribution of arguments. Therefore, I shall attack doctrines and not their assertion by particular authors in particular contexts. It may be felt that all those I attack are straw-men, but the reader familiar with the current literature on this subject will agree, I think, that these straw-men live, breathe, and publish. Having done some critical analysis I hope then to offer the reader a plausible explication of just what one could reasonably mean by claiming that geometry is conventional. In particular I want to examine, all too briefly, the relationship between the claims that geometry is conventional and that *all* physical theory is conventional. I do not think it accidental that geometry has proven a particularly fascinating theory, when questions of arbitrariness and conventionality arise. I hope to show why geometry is special in this way, and to hint at to what extent our understanding of the nature of geometric theories throws light on our general understanding of the epistemology of physical theories.

III

Let us begin by examining a number of possible views as to just what it means to say that geometry is conventional.[3]

[3] For critical analysis, in detail, of many of the arguments noted below see G. Massey, *The Philosophy of Space*, Princeton University Ph. D. Thesis (unpublished). I am indebted to this thesis throughout Sect. III. It also provides a useful bibliography for the sources of the arguments I consider in this section.

Thesis A

The doctrine of conventionality can only be understood by distinguishing between pure and applied geometry. Pure geometries are, indeed, conventional. Applied geometry, or physical geometry, is an empirical theory and not conventional; or, certainly, no more conventional than any other physical theory.

This line has several variants, depending upon what is meant by "pure geometry." I think we can see in each case that the explication fails for two reasons: (1) the "pure" geometry is in no sense conventional in any of the variants of this line of attack; (2) in none of the variations are we given any understanding of the *special* conventionality of geometry as compared with other physical theories. Yet is it clear that geometry has been thought to be, in some sense, "more" conventional than other theories, and whether this claim is true or not, it is incumbent upon us to explicate just *why* such brilliant men as Poincaré and Reichenbach have believed this to be so.

Thesis Variant A1

Pure geometry is geometry with all the terms in the assertion of the theory dis-interpreted.

But then pure geometry consists of meaningless strings of symbols. Meaningless strings of symbols are neither true nor false, not even conventionally true or false! To call such strings of symbols "conventional" is absurd, for only assertions which can be true or false can be conventionally true or false.

Thesis Variant A2

Pure geometry consists of the assertions of geometry with only the *non-logical* terms dis-interpreted. The logical terms retain their ordinary significance.

If the geometry which is dis-interpreted is one of the usual geometries, then this dis-interpreted "pure" geometry will be a consistent set of sentence forms. Hence, the set will be satisfiable and have a model. But still, there is no sense in attributing truth or falsehood to such *forms* of sentences, which are all that remain after dis-interpretation of the descriptive vocabulary, and hence no sense in speaking of this version of "pure" geometry as conventional. Consistency, satisfiability, etc.—these are the properties attributable to such forms, not truth, falsity, or conventionality.

Thesis Variant A3

Pure geometry is *hypothetical* geometry. There is no question of dis-interpretation in this case. For a given formalized (i.e., axiomatized) geometry, the associated pure geometry consists of all sentences of the following form: If (conjunction of the axioms of the geometry), then (any theorem of the geometry).

But, once again, there is no question of this "pure" geometry being conventional—at least in the sense with which we are concerned. Because, if the original geometry was a sound deductive theory, the hypothetical geometry consists simply in logical truths. Let us put this slightly differently: There is no question of a conventionality of *geometry* being involved in such hypothetical geometries. The sentences of such a theory are no more and no less conventional than any other truths of logic. In any case, there is no difference here between geometry and any other formalized theory, since hypothetical biology or hypothetical chemistry is just as much a collection of truths of logic as is hypothetical geometry.

In none of the variants above is anything provided which would make clear to us the nature of claims such as that of Poincaré: that geometry, in some *special* sense is a theory to be adopted by convention.

Thesis B

The conventionality of geometry is due to an isomorphism (or, perhaps, some weaker form of mapping) between either the fully or the partially dis-interpreted calculi of different geometries.

Here we have two variants to consider in turn.

Thesis Variant B1

The isomorphism (or weaker mapping) is between the calculi of the alternative geometries fully dis-interpreted, i.e., with even the logical terms dis-interpreted.

Clearly the existence of any kind of mapping between such meaningless strings of symbols is totally irrelevant to the consideration of the epistemic status of genuine meaningful theories.

Thesis Variant B2

The isomorphism (or weaker mapping) is between the partially dis-interpreted calculi of the alternative geometries, i.e., between what remains of the theories with only their descriptive, non-logical vocabularies dis-interpreted.

The existence of such mappings, especially of a one-to-one, term-by-term "translation" of one partially dis-interpreted calculus into another, is of serious interest for it can serve as one means of establishing a relative consistency proof between two theories. This is a well-known technique, and was of importance in the early days of non-Euclidean geometry. Since doubts were frequently cast upon the consistency of such geometries, anything which could serve as a proof of their consistency relative to the Euclidean theory was of value. A mapping of this sort was presented by Poincaré in one of his relative consistency proofs for non-Euclidean geometry. Here the infinite geodesics of a Lobachevskian space were mapped into chords of a finite Euclidean sphere. By introducing an appropriate mapping of relations, Poincaré was able to map Lobachevskian geometry into a *fragment* of Euclidean geometry. Clearly, a very satisfactory refutation of the claim that non-Euclidean geometries are inconsistent. But this does not show that such mappings are of any interest in trying to decide whether the choice between two geometries is in any way conventional. All that is left after the partial dis-interpretation of a theory is its purely logical structure, and any relationships between such sets of sentence forms can cast light only upon the logical features of the original theories. The existence or nonexistence of such a relationship cannot affect our epistemological views except in such trivial senses as this: if we find the theory inconsistent, we will no longer worry about its empirical support.

It would seem that those who look for the meaning of "conventional" as applied to geometry in the existence of such mappings are simply confusing two different arguments and two different examples presented by Poincaré: (1) his argument concerning the conventionality of geometry, backed up by his two-dimensional parable; and, (2) his proof of the relative consistency of non-Euclidean geometries by means of a mapping of Lobachevskian geometry into a fragment of Euclidean geometry.

Thesis C

The conventionality of geometry is due to the fact that the object of geometric study, space, is a continuous manifold.

The next group of attempts to explicate the meaning of "conventional" as applied to geometries finds the roots of this doctrine in a particular feature of the subject matter of gometry, that is, in the continuity of space. At first sight, this particular feature of space would not seem to have any relevance at all to the problems with which we are concerned. I believe that this first intuition is correct. But since, quite recently, arguments have been presented to the effect that such considerations are crucial, and since there appears to be a remark suggesting this theme in Riemann's famous paper on the physical possibility of non-Euclidean geometries,[4] we will give

---

[4] B. Riemann, "On the Hypotheses which Lie at the Foundations of Geometry" in D. Smith (tr.), *A Source Book in Mathematics*, vol. 2 (New York, Dover, 1959), pp. 413, 424–425. The relevant passages are also quoted in A. Grünbaum, *Philosophical Problems of Space and Time* (New York, Alfred A. Knopf, 1963), pp. 8–9.

these arguments brief consideration. Once again, the argument takes on several variant forms, not always adequately distinguished.

Thesis Variant C1

The denseness (continuity) of the manifold of space allows many different metric functions to be defined on its points. It is these different definitions of the metric which give rise to the different geometries which are, hence, conventionally adopted. That is, space allows you to pick your metric function in different ways, and hence, to adopt different geometries.

As a first point we should note that even discrete spaces allow the choice of alternative metrics, i.e., of more than one function on pairs of points of the space such that: (1) $m(x,y) = m(y,x)$; (2) $m(x,x) = 0$; (3) $m(x,y) \geqslant 0$; and (4) $m(x,y) \leqslant m(x,z) + m(z,y)$. Hence neither the denseness nor continuity of space are relevant to this argument.

What, then, is left to the thesis? Simply the following: If you change the meaning of "distance between points x and y" then you will discover different sentences containing that phrase to be true than those which were true given the old meaning of the phrase. But is this supposed to come as a surprise? Or to be a feature peculiar to geometry? If "lion" meant what "tiger" now means, then "lions" would have stripes, but zoology is hardly a matter of convention.

We shall later see that my own analysis of the meaning of "conventional" as applied to geometries will give rise to doubts as to whether trivial semantic conventionality such as we have just described, is as easily distinguished from more significant kinds of conventionality as might be thought—or as easily disposed of as I have just disposed of it. But this will come at the end of a long story, and not as the result of such trivial considerations as the thesis we are now discussing presents.

Thesis Variant C2

The denseness (continuity) of the manifold of space prevents our defining the metric solely in terms of the topological, i.e., ordering, relations among the points of the space. This means that the metric, hence the Euclidean or non-Euclidean nature of the geometry, is not "fixed" by the topology. This is what is meant by the claim that the choice of geometries is conventional.

In this argument a genuine distinction is noted between dense and discrete spaces. In a discrete space we can define the distance of a point from a chosen origin simply in terms of its ordinal place in the series of points beginning at that origin. In a dense space, the absence of a "nextness" relation among the points clearly makes such a "topologically defined" metric impossible.

What is hard to understand is why this feature of dense spaces has been thought relevant to the issues in question. The argument seems only to show the following: one can (1) fix a topology for a dense space, (2) change one's meaning of "distance" without modifying the already specified topology, and (3) cannot define one's meaning for "distance" solely in terms of the ordering of points in the space as specified by the topology. But, once again, what we appear to have is merely trivial semantic conventionality, with a few interesting but quite irrelevant facts about the properties of dense manifolds superadded.

Neither version of Thesis C, then, seems to show that the denseness or continuity of space is in any way a relevant consideration in trying to determine in what, if any, peculiar way geometry is conventional.

Thesis D

The conventionality of geometry is simply this: If we choose to assign "distances" between pairs of points differently, then we shall adopt different laws about "distance." Such alternative laws about "distance" are alternative geometries, so it is open to us which "distances" we assign and, hence, which geometry we adopt.

This argument, of course, is simply the bald assertion of trivial semantic conventionality, but now separated from the irrelevancies of Thesis C. It is not truly a thesis designed to illuminate the special status of geometry with regard to conventionalism, but rather to *deny* geometry any special place in this regard among theories. For, as we have noted above, in the sense of trivial semantic conventionality, every theory is conventional. I have noted this argument as a separate thesis, not in order to discuss it separately, but only to keep this position fresh in the reader's mind. For, as we have already mentioned, after a long tale, full of important digressions, we shall ultimately find ourselves faced with the following question: Can such "trivial" renamings of properties and relations be as easily dismissed as we have made them out to be? As we shall see, modification of physical theories by "trivial" renaming is not as simply distinguishable from genuine scientific revisions as might at first appear. This is a familiar conclusion to those acquainted with critical analysis of empiricist dogma, but we shall leave our examination of it to the end of our story, where it properly belongs.

IV

Having disposed of implausible explications of Poincaré's thesis, it is now incumbent on me to offer a sensible account of just what he might have meant when he said that geometry is merely a matter of convention. How to proceed is made perfectly clear by a consideration of Poincaré's original two-dimensional parable, which he offered to convince the reader of his claim.

Poincaré asks us to imagine a finite, two-dimensional, Euclidean disk, inhabited by a community of two-dimensional scientists.[5] These scientists, having read their two-dimensional Riemann, set about to determine the geometry of their world, using rigid rods to triangulate their home in the usual way. Unbeknownst to them, we, in our three-dimensional malevolence, have imposed upon their world a temperature gradient. The temperature reaches its maximum at the center of the disk and diminishes, with radial symmetry, to absolute zero at the bounding circle. Furthermore, we are to assume that the two-dimensional's measuring rods all expand and contract with temperature at the same rate, irrespective of their construction, and that they all shrink to zero length as the temperature reaches absolute zero. With the appropriate rate of diminution of temperature from center to periphery, and the appropriate laws of thermal expansion, Poincaré shows, we can deceive the flat-landers into thinking that they are the inhabitants of an infinite, Lobachevskian world of constant negative curvature.

As a slight complication, Poincaré allows the two-dimensional creatures the use of optical experiments in the geodesy of their world. To keep them from discovering the deception we have plotted, we need only allow for the modification of the index of refraction of their two-dimensional optical medium. Appropriately modified, the distortions introduced will lead them once more to conclude erroneously that they live in a world of infinite extent and constant, negative curvature.

The significance of the parable becomes clearer when we postulate, as a denizen of this two-dimensional world, a two-dimensional scientist of sufficient genius to speculate that his world is not, as other scientists have imagined, infinite and Lobachevskian, but rather Euclidean and finite, and cursed with just such distorting physical fields as we have described above. How is the Flatland Physical Society to decide between the traditional view and our scientist's novel proposals? Without access to our three-dimensional world they simply cannot make any decision based on observation or experiment. Their only rational policy, therefore, is to allow everyone his own consistent *choice*: either believe the world is infinite and Lobachevskian, with no peculiar distorting fields, or believe the

[5] H. Poincaré, "Space and Geometry," *op. cit.*, pp. 65–68.

world Euclidean and finite, but cursed with the distorting influences we have described.

The next step is clear. Project what we have just described into our own three-dimensional world. Here there is no ideal, Euclidean four-dimensional space in which our three-dimensional world is embedded. At least, we have no reason whatever to believe in the existence of such a higher dimensional embedding world. In this three-dimensional world, our world, we are faced with the same undetermined choices as faced the flat-landers, when our task is to propose a geometrical hypothesis to account for any collection of metric experiments. Therefore, the choice of geometry by which we describe the world is a matter of convention.

As we have noted, Poincaré believed that the simplicity of the Euclidean choice would always make it preferred, whereas Reichenbach, taking account of the results of general relativity, allowed for the possibility that we might choose a non-Euclidean geometry in order to postulate a "descriptively simpler" total theory of geometry plus remaining physical theory.[6] Other than this, Reichenbach's account closely parallels Poincaré's.

In passing, we might note that Reichenbach unfortunately described the modifications to our physical theory necessary to preserve an apparently refuted geometry as the postulation of "universal forces." Clearly, "universal stretching field" would be a more apt title in the case of those fields necessary to "stretch" or "shrink" rigid rods upon translation. "Universal optical medium distorting field" would be an apt (if too-long) name for that field which must be postulated to "compensate" appropriately for getting the right results from optical experiments. Reichenbach also failed to realize that we could do without "universal" fields by postulating multiple "differential" fields whose *effects* (on rigid rods and light-rays) added up to the effect which would have been produced by the appropriate "universal" distortion.

V

By this time the name of Pierre Duhem will probably be ringing loudly in the reader's ears. Isn't the analysis of the conventionality of geometry which I have offered, following Poincaré and Reichenbach, simply an instance of the famous Duhem thesis regarding crucial experiments? We are presented with experimental results

[6] H. Poincaré, "Non-Euclidean Geometries," *op. cit.*, p. 50. H. Reichenbach, *op. cit.*, p. 34.

which apparently refute a particular geometric hypothesis, say, that space is Euclidean. Rather than reject our original hypothesis in the face of this recalcitrant evidence, we make substantial modifications in other portions of our physical theory by postulating universal distortions. These modifications serve to save our original geometric hypothesis, for the original geometry plus the *new* remaining physical theory is as compatible with the experimental results as would be a novel geometry combined with the *old* remaining physical theory.

But the situation is not quite as simple as this. For the two combinations, Euclidean geometry and non-Euclidean geometry each with its appropriate auxiliary physical theory, are not simply supposed to be both compatible with *one* experimental result, or with a limited set of experimental results. The claim is, rather, that *any* experimental result compatible with one of these combinations is compatible with the other. Whereas the ordinary alternative Duhemian combinations are "equivalent" relative to some one experimental result, or to some limited class of such observations, these "super-Duhemian" alternative total theories are supposed to be equivalent with regard to any possible experimental test. This is the point of Reichenbach's claim that there is nothing to choose between the two total theories except matters of "descriptive simplicity," and this is the full import of the claim that a choice one way or the other is merely a matter of convention. This "super-Duhemian" aspect of the thesis of geometric conventionalism has been emphasized by Adolf Grünbaum.[7] He has drawn from his observations, however, rather strong conclusions relating to questions of the meaning of the terms of the geometric theories. We shall scrutinize these conclusions further on in this paper.

This, then, is our analysis of the meaning of the claim that one's choice of geometry is a matter of convention: For any given geometric theory, there will always be alternative geometric theories such that, if sufficient modifications are made in the portions of physical theory which remain after geometry is left out, each of the alternative geometries will be compatible with any experimental results which were claimed to have supported the original theory. There will be nothing to decide one's choice between the alternative theories but matters of elegance or convenience, called "descriptive simplicity" by Reichenbach. Hence, one's choice of which geometry is said to describe the world is purely a matter of convention.

[7] A. Grünbaum, "Law and Convention in Physical Theory" in H. Feigl and G. Maxwell (eds.), *Current Issues in the Philosophy of Science* (New York; Holt, Reinhart and Winston, 1961), see esp. his "Rejoinder to Feyerabend," p. 164.

VI

The analysis of the meaning of the claim of geometric conventionalism must not stop at this point, for the last paragraph immediately raises a number of crucial and disturbing questions. The remainder of this paper, its most important sections, will be devoted to raising three such questions. For two I shall try to provide at least partial answers. The third is sufficiently deep and puzzling to discourage any attempt at providing an answer brief enough for a paper such as this. I shall only try to indicate where the problems lie in seeking a solution to it, and will do nothing to indicate a plausible answer. Let me begin by asking all three questions first. Then I shall consider each of them in turn.

(1) The conventionalist claim makes crucial use of the notion of two theoretical combinations each "saving the observable facts." But can we make coherent sense of this notion, such that it becomes plausible to say that two mutually incompatible theories can still be said to agree on "all observable consequences"?

(2) Has our analysis succeeded in showing why geometry is in some way *special* with regard to the conventionalist thesis? Can't the conventionalist claim, as I have explained it, be extended to any physical theory whatever? If "conventional" means what I have said it means, then isn't every theory conventional?

(3) What is the deeper import of such "super-Duhemian" alternative total theories? Does the existence of such alternatives have important consequences for our understanding of the nature of the meaning of theoretical terms? Does it have important consequences for our views about the ontology of theories?

Each of these questions in turn will now be considered.

VII

First Question. Can we give a meaning to the notion of "saving the observational consequences" sufficiently clear to make the conventionalist doctrine intelligible? To see why the answer has been thought to be affirmative we must cease dealing with our problem in total philosophical abstraction, and proceed to examine the manner in which problems of physical geometry are treated in actual scientific practice. Of crucial importance is the introduction in such scientific discussions, mostly in the context of the exposition of the theory of general relativity, of the concept of the idealized measuring device. In every such discussion reference is made to such entities as ideal rigid rods—rods immune to the distorting physical influences to which ordinary measuring rods are subject; ideal clocks—once again immune to the "ordinary" disturbances to which all ordinary clocks are prey; and ideal light waves traveling in perfect vacuua. It is always the behavior of such ideal entities, never to be found in the distinctly nonideal physical world in which we actually live, which are described by the author as revealing the structure of the space or space-time of the universe being described.

The second feature to be noticed in the scientific discussions of the nature of space or space-time is the assumed "pure theoreticalness" of the spatial or spatio-temporal entities themselves. We do not, according to the texts, ever observe spatial intervals themselves, but only the behavior of transported ideal rigid rods, or the comparable paths of intersecting rays of light. Nor do we ever "observe directly" temporal intervals, but only the behavior of idealized physical clocks synchronized with events and with each other.

To discover the nature of space, then, we make "observations" on the behavior of idealized physical objects such as rigid-rods, clocks, and light-rays. But for such observations to supply us with an insight into the nature of space, assumptions must be made as to just how the nature of space "affects" the behavior of the idealized measuring instruments. Such assumptions are, of course, just the auxiliary physical theory which must be modified, according to the conventionalist view, if we are to change geometries while "saving the phenomena." The laws connecting the structure of space with the behavior of idealized physical objects in space are familiar to us as Reichenbach's *correspondence rules*. But two features distinguish correspondence rules as we have just described them from Reichenbach's original view of them:

(1) They connect, not space and observations, in the ordinary sense of "observations," but rather the nature of space and the behavior of idealized measuring instruments. These measuring instruments, so far from being "phenomenal" objects in the philosopher's sense, are not even actual *physical* objects, but are only the idealized fantasies of theoretical discourse.

(2) While Reichenbach took correspondence rules as definitional, we have as yet made no commitment as to the "analytic" or "synthetic" nature of these laws, but only treated them, so far, as on a par with any other generalizations of physical theory.

Item (1) above is important for our present discussion. We shall return to (2) shortly. What (1) shows us is that the plausibility of the conventionalist position does not depend upon a general epistemological notion of "observation basis," nor upon any of the other views generally associated with positivist or phenomenalist doctrine. Rather, it depends upon two features characteristic of geometry as opposed to physical theory in general:

(1) Our intuitions that spatial (or spatio-temporal) intervals are truly non-observable, in *any* plausible sense of "observable."

(2) The nature of the scientific discussion of physical geometry with its postulation of idealized measuring instruments, and where, *within this scientific context*, the notion of observable phenomenon is restricted to that of the "in principal observable" behaviour of these idealized physical entities.

We may now summarize what "saving the phenomena" amounts to in conventionalist doctrine: (1) You can't observe spatial or spatiotemporal intervals directly. (2) Any knowledge you have of them is gained by "observation" of idealized measuring instruments. (3) But such observation indicates the nature of space or space-time only if the nature of space or space-time is connected by some general laws with the behavior of the idealized physical objects. (4) Hence, by suitably changing one's views as to just what laws do connect the structure of space or space-time with the behavior of idealized physical objects, one can postulate alternative geometries and yet at the same time predict the same behavior for the ideal measuring instruments in each of the alternative cases.

Second Question. Why does the conventionalist doctrine seem plausible in the case of geometry, but not in the case of other physical theories? We have already supplied part of the answer when we discussed question one above. It is that the entities described by geometric theories have a "hyper-theoretical" quality to them. We might plausibly claim to "see" molecules, or even to "sense directly" electric and magnetic fields, but our intuition (rightly or wrongly) suggests something almost absurd in the notion of "seeing" the spatial interval between two spatial points. Not seeing two objects located at these points, mind you, nor some object whose end points are located at these points, but seeing the interval itself. Perhaps this intuition is deceptive. Indeed, it might plausibly be asserted that the whole claim is one framed in concepts which would be seen to be pointless and incoherent from a reasonable philosophical epistemology. But the intuition is one which we have, nonetheless. If this does not show that geometry is conventional in a way in which other theories are not, it may be at least part of an explanation as to why Poincaré *felt* that geometry was conventional, and yet never asserted such conventionality for run-of-the-mill physical theories.

But there is another, stronger influence behind the ability of geometry to suggest conventionalist doctrine. This is the fact that geometry was the first *formalized* physical theory and, even at this date, is the only physical theory so neatly and elegantly formalized.

How does formalization affect our epistemological views about geometry? First, there is the historical influence. Because geometry was formalized so early, and because the formalization rested upon such "self-evidently true" axioms, it was geometry which suggested itself to everyone from Aristotle to Kant as the paradigmatically *a priori* physical theory. We have already seen how this fact motivated Poincaré's conventionalist position.

But a second feature of formalization is even more important, today, in supporting the conventionalist position. For *if* we restrict our attention to the actual scientific context, with its fairyland of ideal rods, clocks, and light-rays, then we find that not only can we *speculate* upon the existence of alternative "super-Duhemian" combinations of geometry and remaining physical theory, but we can actually *construct*, with some ease, such alternative "super-Duhemian" pairs. Because of the formalization of geometric theory, and of the theory connecting the structure of space to the behavior of idealized physical objects, it becomes possible to describe the changes necessary in the latter theory to compensate for any proposed changes in the former. A prime example of this is the reformulation of general relativity to allow for a Euclidean space-time, while at the same time preserving such predictive consequences of the theory as the advance in perihelion of planetary orbits, the bending of light-rays near massive objects, the slowing down of clocks in gravitational fields, etc.[8] Only in the case of geometry, among all physical theories, do we have the possibility, as science now stands, of actually displaying the alternative total theories postulated to exist by conventionalist doctrine.

Before going on, I would first like to make a digression which I think is of some importance. It is crucial to notice the protasis of the conditional second sentence in the paragraph above. One can easily construct such alternative total theories *only* in the scientific context

---

[8] See, for example, Peter Havas, "Foundation Problems in General Relativity" in M. Bunge (ed.), *Delaware Seminar in the Foundations of Physics* (New York, Springer-Verlag, 1967), pp. 140–141, and the reference therein cited.

with its idealized entities. Einstein pointed this fact out. It disturbed Reichenbach and has recently been touched on again by Putnam in his critique of Grünbaum's conventionalist position.[9] If the measuring rods we are talking about are *actual* measuring rods, then in order to be sure that our alternative total theories have the same predictive consequences we should have to assure that the alternative theories not only compensate for changed views about the basic structure of space, but for all the incumbent changes in the disturbing influences to which real measuring rods are subject. Since the prediction of these changes depends upon knowing the laws of nature, and since each law presupposes an underlying geometry, it is not obvious just how the alternative portion of theory outside of geometry is to be constructed. Of course it would be unreasonable to say that this is impossible. In fact, its possibility seems evident. But it is only because we restrict our attention to ideal objects, that we can so easily, systematically, and elegantly construct the alternative auxiliary physical theories.

Once again we see the importance for the conventionalist position of sticking to the way geometry is actually handled in scientific contexts. It is only in this context that we can clearly distinguish the observable from the unobservable. The former is the behavior of ideal objects, the latter the structure of space-time. It is only in this context that we can actually construct the alternative "super-Duhemian" theories postulated to exist by conventionalist doctrine. It is now somewhat clearer why geometry is a special case for the conventionalist view of physical theories.

Third Question. At this point we must abandon the relative clarity of the preceding two sections. For our third question allows of no obvious or satisfactory answer—at least at this stage of philosophy. I shall be satisfied if I bring out for the reader just part of the perplexity which lies hidden in our apparently innocent concern for the meanings of geometric terms. In a sense this can all be considered a footnote to Poincaré's all too casual remark about the axioms of geometry fixing the meanings of geometric terms. It is also a return to the question of the "analyticity" of Reichenbach's correspondence rules, a question I noted earlier and promised to give some attention.

Let us approach the problem as follows. Suppose that we are presented, as physicists, with a geometric theory and its associated

[9] Albert Einstein, "Remarks" in P. Schilpp (ed.), *Albert Einstein, Philosopher-Scientist* (New York, Tudor, 1949), pp. 676–678. See also H. Reichenbach, *op. cit.*, pp. 20–24 and Hilary Putnam, "An Examination of Grünbaum's Philosophy of Geometry" in B. Baumrin (ed.), *Delaware Seminar in the Philosophy of Science*, vol. II (New York, Interscience, 1963), p. 246.

"correspondence theory," the latter connecting the structure of space-time to the behavior of idealized rods, clocks, and light-rays. We are then challenged, as conventionalists, to supply our vaunted alternative total theory. We proceed to do so in the manner described above, choosing our geometry "according to taste" and modifying our auxiliary physical theory to arrive at the same conclusions regarding the behavior of the idealized physical objects as were inferable from the original total theory.

But now our interlocutor challenges us. "You have obtained your end," he asserts, "by sheer fraud. What you have done is not to supply a novel geometry at all. All that you have done is to take the old geometric terms and assign them new meanings. That you come out with new true sentences containing those terms is hardly a surprise. Anyone who has thought about trivial semantic conventionality for a moment would expect that." But why, we ask, does he believe that we have changed the meanings of the geometric terms? "Take the term 'geodesic,' for example," he replies. "I, and you before you played your silly game, *meant* by 'geodesic' the path along which an idealized light-ray in vacuum travels. Now your novel theory informs me that light-rays do not travel along geodesics at all. It is plain, we simply do not any longer mean the same thing by 'geodesic'."

By itself this argument is hardly convincing. But when he goes on to point out that many of the most basic sentences which connected geometric to nongeometric terms in our auxiliary theory have been changed, due to the radical changes in the auxiliary theory incumbent upon picking a new geometry, we begin to see his point. "After all," he continues, "what on earth *gives* meaning to the geometric terms, terms designating purely unobservable entities, aside from their appearance in general sentences in which terms designating observables also appear? But this is just their appearance in the laws of the old auxiliary theory which you now reject. To change *all* of these laws is simply to change the meanings of the geometric terms therein. Hence, an *apparently* new geometry appears. It is no surprise that your new total theory predicts the same behavior for rods, clocks, and light-rays as your old. They are the same theory, written in two different ways. One might as well be surprised that Newton's *Principia* has the same observational consequences in its Latin as in its English versions."[10]

[10] For a proposal that such "super-Duhemian" revision is merely trivial semantic conventionality see Arthur Eddington, *Space, Time and Gravitation* (Cambridge, University Press, 1953), p. 9. See also A. Grünbaum, *op. cit.*, pp. 219–229.

The radical thesis has recently been proposed that *any* change in a theory in which a term appears changes the meaning of that term in the language.[11] The usual reply to this reduction to absurdity of the concept of meaning is to claim that if most of, or the most significant parts of, the "cluster" of laws in which the term appears remains unchanged, then it is only reasonable to say that the meaning of the term remains invariant. This is certainly the case if the laws remaining unchanged are those "basic" ones more essential for fixing the meaning of the term. But here the problem seems deeper, since a great many of the most basic laws containing the geometric terms are changed in adopting our new total theory. These include those laws of geometry which contain only geometric terms as well as those containing both geometric and nongeometric ones and which serve to connect geometry to the remaining physical theory.[12] From this point of view the postulation of "universal distorting fields" really amounts to a change in "definition" of the geometric terms, say, "distance" and "geodesic."

Does conventionalism, as we have described it, then, simply reduce to trivial semantic conventionality? Not quite. If we have reached trivial semantic conventionality, we have done so by a very circuitous route, indeed. I am tempted to take refuge in the claim that as far as philosophy is concerned, if the routes differ enough, the end point really cannot be said to be the same. For how you got there is as important as where you arrived.

More seriously, I believe that what we have discovered is not that geometric conventionalism is merely trivial semantic conventionality warmed over, but rather that questions of the meaning of terms, when the terms are at the most abstract theoretical level as are those of geometry, are simply not amenable to our present ways of talking about meaning. It would appear, naively, that there are only two alternatives: (1) the "two" alternative total theories are merely the same theory reworded, or (2) the two alternative theories are incompatible and distinct physical theories which happen to have the same observational consequences. But is the distinction between (1)

---

[11] See Paul Feyerabend, "Explanation, Reduction, and Empiricism" in H. Feigl and G. Maxwell (eds.), *Minnesota Studies in the Philosophy of Science*, Vol. III (Minneapolis, University of Minnesota Press, 1962). Also H. Putnam, "How Not to Talk About Meaning" in *Boston Studies in the Philosophy of Science*, Vol. II, ed. by M. Wartofsky (Dordrecht, D. Reidel, 1963), and Arthur Fine, "Consistency, Derivability and Scientific Change," *The Journal of Philosophy*, vol. 69 (1967), pp. 231–240.

[12] In changing from, say, Euclidean to Lobachevskian geometry, not *all* sentences of the geometry are changed, but surely enough, and those central enough, to make the "most of the cluster remaining the same" argument implausible.

and (2) really sufficiently clear for a decision to be made one way or the other? Perhaps what we have provided is additional ammunition to those who would bury "meaning" along with other dead, incoherent philosophical concepts. But this is a much deeper question than any which I wish to discuss here.[13]

Naturally, along with these deep questions on the meaning of geometric terms go equally deep puzzles about the ontology of geometric theories. Once again I can only hope to reveal the perplexity, not assuage it. Considerations similar to those we have been discussing in the case of the meaning of geometric terms have led some philosophers and scientists to believe that geometry has no ontology at all, properly speaking. For, they argue, if the structure of the putative geometric realm is merely a matter of convention, how can one plausibly believe in such an "independently existing" realm of entities at all? Surely if there are geometric entities, what they are like is no matter of convention. Indeed, if the choice of which geometry is "true" or "false" is merely conventional, why say that, properly speaking, geometric theories are true or false at all? But clearly we are getting beyond the shores I hoped to explore, and into very deep waters, indeed!

Even more fascinating, and even more to be avoided at the present time, is a question which has probably occurred to the reader more than once: Granted that conventionalist doctrines are most plausibly motivated and most easily stated in the case of geometry, isn't it still the case that the same problems reappear, perhaps more darkly, in the case of any physical theory? Shouldn't we say, finally, that the thesis of conventionalism for physical theory in general appears all too plausible?

University of Michigan

---

[13] We have indeed returned to the question as to whether Reichenbach's correspondence rules are "analytic" or "synthetic," but only to suggest that there may simply be no point in asking this question in the first place, even if we "emasculate" the correspondence rules to make them hold between geometrical entities and ideal physical objects rather than between geometric entities and "observations" (as positivists originally hoped).

IV
What are Physical Theories About?
MARIO BUNGE

AFTER forty years of complacency, quantum physicists are starting to do some soul-searching. In an increasing number of papers and lectures they point out the reason for their dissatisfaction, namely that there are hardly any acceptable profound theories accounting for the many quaint "elementary particles" and atomic nuclei. They confess that, though they can compute a number of quantities, they are at a loss to understand what is being calculated, for only a comprehensive and deep theory can yield a satisfactory understanding—and such a conceptual system is precisely what is wanting. In short, it is felt that quantum mechanics and quantum electrodynamics, for all their astounding coverage, are insufficient: that new theories should be built to solve the mounting heap of open problems. And, as is normal in such crises, the strategy problem arises: Which way should physics go, i.e., what kind of theories should one attempt to build?

As a physicist I could not care less which way fundamental physics goes as long as it keeps going. In particular, I do not care whether physics remains basically probabilistic or else becomes even more radically stochastic (by replacing all extant parameters and dispersion-free variables by random variables) or, what seems unlikely, turns around and goes deterministic in the classical sense. But as a philosopher I do care about the following: (a) that physics remains *physical*—i.e., concerned with physical objects rather than with observers and their mental states (e.g., expectations and uncertainties); (b) that physics goes ever *deeper*, from phenomenological (black box) to semiphenomenological to mechanism theories; and (c) that physics becomes ever more *cogent*—that its theories be formulated in more consistent and better organized ways. Yet the prospect of an advance in such a direction is by no means assured, for a fashionable thing to do in fundamental physics is to toy with nonphysical (particularly psychological) ideas, to prefer the black box to the translucid box, and to tolerate inconsistencies as long as one can get some computations done. Indeed, there is not even assurance that fundamental physics may go beyond its present stage—among other reasons because of the rather widespread belief that we have already gone all the way we humanly could. Obviously, if one believes that any present theory is an ultimate, then he will make no effort to go beyond it.

As in many a religious crisis, some physicists believe in the beyond while others do not. But, whether or not they do, their attitude is largely a matter of faith supported or undermined by past experience as well as by some philosophy or other. The believer has faith in scientific progress while the disbeliever either has faith in the ultimate character of current theory, or he doubts the possibility of overcoming its limitations. Mine is a conditional belief. I think the beyond—that is, a thoroughly physical quantum theory even richer and more cogent than the present one—is badly needed. But at the same time I do not believe such a theory, or rather set of theories, will be invented unless people want to. And, even if they do wish to go ahead, they are bound to meet formidable obstacles: physical, mathematical, and philosophical ones. The main technical difficulty would seem to be one of getting rid of classical analogies (still infesting the standard presentations of the quantum theories) and conceiving instead unheard-of ideas matching the properties and laws of those odd things called "elementary particles." This may well require the assistance of new mathematical tools, just as classical mechanics and classical field theories did in their time. And the main philosophical obstacle is the current confusion and uncertainty concerning the referents of fundamental physical theories, i.e., the kind of things they are about.

The purpose of this paper is to examine the latter question, i.e., to try to identify the kind of things physical theory is concerned with. This, at least, is not a matter of faith but one of logical analysis and argument. And unless we clear it up we will not know whether a beyond is conceivable and, if so, which way it lies. For, if the answer is that the present quantum theories are about object-apparatus-observer blocks that cannot be further analyzed (Bohr's "essential wholeness of a proper quantum phenomenon"[1]), then obviously there is no beyond. But if, on the other hand, physics must introduce the observer's mind into the picture of the world (as Wigner[2] has proposed), then there is hope for progress as long as physics joins psychology. And if, finally, physics is about things that are presumed

---

[1] Niels Bohr, *Atomic Physics and Human Knowledge* (New York, Wiley, 1958), pp. 72 and *passim*.

[2] E. P. Wigner, "Remarks on the Mind–Body Question" in *The Scientist Speculates*, ed. by I. J. Good (London, Heinemann, 1962).

to be out there, then its task remains the traditional one of getting to know more and more about them, rather than either proclaiming final victory or turning inward, to the study of the self.

I. The Interpretation Problem

1. *The Referent*.

That which a construct (concept, proposition, or theory) is about, or stands for, or refers to, or rather is intended to refer to, is called the (intended) *referent* of the construct. The referent of a construct may be a single object or any number of objects; it may be perceptible or imperceptible; it may be real, presumably real, or imaginary, and so forth. In any case the referent of a construct is a collection of items and is therefore also called the (intended) *reference class* of the construct. For example, the reference class of "The ether is elastic" is now believed to be empty, while the one of "The earth spins" is a singleton; the extension of "The sun revolves around the earth" is a pair, and the one of "All quarks have a fractionary electric charge" is a set of an unknown cardinality.

Some reference classes are *homogeneous*, i.e., composed of elements of a single kind—e.g., deuterium atoms. Other reference classes are *inhomogeneous*, i.e., composed of elements of distinct kinds, such as protons and synchrotrons, or atoms and external fields. A reference class composed of pairs of entities of the kinds A and B may be construed as the cartesian product $A \times B$ of the corresponding sets. In general, a reference class is a set of n-tuples, i.e., a cartesian product of n (not necessarily different) sets. A construct will be said to *refer partially* to a class A just in case the set A occurs in the reference class of the construct, either as a subset or as a cartesian factor of the reference class. (Incidentally, an n-tuple need not be construed in the standard way, i.e., as a set: it may alternatively be construed as an individual, and consequently be introduced as a primitive concept rather than as a defined one. (This is in fact how Bourbaki introduces the concept of ordered pair.) With this construal, reference classes are sets of individuals rather than sets of sets, which as it should be, for the referent of a scientific construct is a thing, not a concept.)

Our problem is to find out the nature of the reference class of a physical theory. In particular, we want to ascertain whether the reference class of a physical theory is made up, if only in part, of cognitive subjects—e.g., observers—or of their mental states. For there is little doubt that some of the expressions occurring in physical writings do refer legitimately, at least partially, to cognitive subjects. Any such expressions will be called *pragmatic expressions*, in contrast to *physical object expressions*, which are free from any reference to cognitive subjects. For example, "The value of the property P for the physical object x equals y" is a physical object sentence (or rather sentence schema), while "Observer z found the value y for the property P of the physical object x" is a pragmatic sentence schema. Clearly, while in the first case reference is made only to a physical system, in the latter case there are two referents: a system and an observer. Or, if preferred, while in the first case it is a question of the *value* of a physical quantity, in the other case it is a question of an *observed value* or empirical estimate of the same quantity, i.e., of its value for a given observer. This difference may look tiny but it is both scientifically and philosophically significant, as will be shown shortly. Suffice now to note that, whereas the preceding physical object sentence has the form: $P(x) = y$, the corresponding pragmatic sentence may be analyzed as $P'(x,z) = y$ or, even better, as: $P'(x,z,t,o) = y$, where t stands for the measurement technique and o for the sequence of operations employed by the observer z in implementing that technique when measuring P on x. We have chosen the new symbol P' to designate the measured property because it stands for something patently different from P: indeed, while the function P is defined on a set X of physical objects, the function P' is defined on the set $X \times Z \times T \times O$ of quadruples physical object-observer-measurement technique-sequence of acts. We shall return to this point in Sect. II.

It is hardly disputable, then, that the language of physics does contain pragmatic sentences. (And also pragmatic metasentences like "Nobody knows whether the quark hypothesis is true.") What is being controverted is the thesis that *all* the nonmathematical sentences occurring in every physical *theory* are pragmatic sentences in the sense that at least one of the components of the reference class of a physical theory is a set of human subjects, such as qualified observers. In other words, what is still a matter of opinion among physicists is the semantical problem of identifying the referent of a physical theory either as a physical system, or as a subject, or as a subject–object synthesis or, finally, as a subject–object pair. In short, what is still controversial is the interpretation of physical symbols and, in particular, the interpretation of the formulas of theoretical physics: are they physical object sentences, or else mental object sentences, or perhaps physico-mental sentences or, finally, partly physical and partly mental object sentences? Before rushing to pick the right answer we ought to know what the interpretation possibilities are.

2. Interpretation: Strict and Adventitious.

However arbitrary theological hermeneutics may be, the interpretation of scientific formulas should not be a matter of choice. To begin with, the interpretation assigned to the basic or undefined symbols of a scientific theory should not render it inconsistent and should turn it true—or, more realistically, maximally true. (For example, it would be wrong to interpret the square of the wave function as a mass density, for this would be inconsistent with the normalization condition.) Second, if a symbol is defined or derived in terms of previously introduced signs, then its meaning should "flow" from the latter rather than being concocted *ad hoc*. (For example, it is wrong to interpret the time derivative of the average of a position coordinate as an average velocity unless the variable can be interpreted as representing a physical position, which is far from obvious in relativistic quantum mechanics.) Third, a strict interpretation of a complex expression should be compatible with the structure of the latter; in particular, if it is claimed that a certain complex symbol concerns a thing of a given kind, then at least one of the constituents of the symbol must be capable of denoting that particular thing. (For example, in order for the wave function to concern both a microsystem and an apparatus, it must depend on variables of the two of them, which is the exception rather than the rule.)

The previous conditions seem obvious yet they are often ignored. The first condition is just pointless in relation with most theory formulations, for it applies only to axiom systems: indeed, it is only in an axiomatic context that the basic/defined dichotomy makes sense. The second condition is violated whenever a defined quantity (or a derived formula) is interpreted in terms that are alien to the defining terms (or to the premisses, as the case may be). This condition is violated, for example, when the entropy of a physical system, computed on the basis of data and assumptions concerning the system itself, is interpreted as a subject's amount of information concerning the system, even though no premises concerning the subject and his fund of knowledge are supplied. As to the third condition, its violation is exemplified by the following typical case. If someone claims that a formula such as "$y = f(x)$" concerns the f-ness (whatever this property may be) of an object x of a certain kind X, such as observed by an observer z of some kind Z, then he is introducing a ghostly variable, namely z. This variable (and the whole set Z) is phony because footless: in order for something to count as a genuine referent it must hold the reference relation to some sign, and in the above case no such symbol corresponding to the alleged observer z occurs in the given expression. (As will be seen in Sect. III, 2, the standard quantum theory of measurement contains such phony variables.)

An interpretation of a (nonformal or descriptive) variable shall be called *strict* if it assigns the variable just one object. If every nonformal symbol in an expression is assigned a strict interpretation, the expression will be said to be interpreted in both a strict and *complete* way. If at least one of the symbols but not all of them are interpreted in a strict way, then the interpretation will be called strict and *partial*. Any interpretation, whether partial or complete, that is not strict will be called *adventitious*. For example, the interpretation of the symbol "$v(x,y)$" as the velocity of a system x is strict and partial; its interpretation as the velocity of a system x relative to a reference frame y is both strict and complete; and its interpretation as the velocity of a system x relative to a frame y as measured by an observer z with the technique t, is adventitious as far as the variables z and t are concerned.

Clearly, a strict and complete interpretation is preferable to either an incomplete or a redundant one: we should neither fail to read some of the components of the meaning of a symbol nor read too much in it. However, not all strict interpretations produce true formulas and not all adventitious interpretations produce false ones. That a strict interpretation can lead to falsity is shown by interpreting, say, the formulas of thermodynamics in terms of subjective probability. That an adventitious interpretation can be true, even trivially so, is equally clear: thus in the example of a function f relating two variables, if f is the right one and the experimenter does his job properly, he will obtain values of f close to the calculated ones. But the experimenter may bungle, in which case the adventitious interpretation becomes false. Also, thus stressing the role of the experimenter may create the impression that the object owes its f-ness to the former, in which case the adventitious interpretation becomes misleading. No such risks are run by strict interpretations. Therefore we must cast a cold eye at adventitious interpretations.

3. *Pragmatic Interpretations.*

Since the birth of operationism there is a strong tendency to construe all linguistic expressions in pragmatic terms. This happens not only in relation with formulas susceptible to empirical tests but

WHAT ARE PHYSICAL THEORIES ABOUT? 67

also in relation with mathematical formulas. This is common practice in the classroom, where it has some didactic virtues, and is the mark of mathematical intuitionism—the mathematical partner of physical operationism. Yet all such pragmatic interpretations of mathematical symbols are adventitious, for a point of mathematics is to abstract from users and circumstances in order to achieve both universality and freedom from commitment to fact. Take, for instance, the asterisk symbol used for complex conjugation. A strict interpretation of the expression "$z*$," where z designates a complex number, is this: "$z*$" means the real part of z minus i times the imaginary part of z. (This rule of designation might be replaced by a definition.) By contrast, a pragmatic interpretation of the same symbol is this: "Anyone presented with the symbol '$z*$' is supposed to reverse the sign of the imaginary part of z." A second pragmatic interpretation of "$z*$" is in the form of a rule or prescription: "To compute $z*$ from z, reverse the sign of the imaginary part of z." A third pragmatic reading of the same symbol would be an instruction capable of being fed into a computer so as to enable it to handle the symbol. Every pragmatic interpretation of a logical or mathematical symbol may be construed as an instruction for handling (e.g., computing) the symbol in an effective way.

Given a mathematical symbol, it can be assigned any number of pragmatic interpretations, according to the user, the circumstances, and the goals (e.g., in connection with different kinds of computers). This plurality of pragmatic interpretations is possible because they are adventitious whenever they concern mathematical symbols: that is, they are not subjected to the internal mathematical laws the symbols themselves satisfy. (Indeed, pure mathematics tells us nothing about users, circumstances, or goals.) It is therefore to be expected that only a subset of the conceivable pragmatic interpretations of a mathematical symbol will be valid. In any case, we need a validity criterion. We shall presently propose one applying to both mathematical and factual symbols.

We stipulate that a pragmatic interpretation of a sign be *valid* just in case there exists a theory containing that sign and such that it provides a ground or reason for the procedure indicated by the pragmatic interpretation. Thus arithmetic provides a ground for (it justifies) the instructions given children for operating an abacus, as well as the instructions fed to a computer in order to find, say, a given power of an integer. A pragmatic interpretation will be called *invalid* just in case it is not valid. Thus interpreting a chemical formula in pragmatic terms involving incantations rather than, say, mixing, stirring, and heating, would be open to the charge of invalidity, for there is no theory countenancing a relation between chemical structure and incantations. Likewise, the interpretation of entropy as a measure of our ignorance is invalid, for it involves the erroneous identification of statistical mechanics with epistemology.[3]

Whether the formulas of theoretical physics may be given valid pragmatic interpretations, remains to be seen (see Sect. II). What is hardly disputable is that pragmatic interpretations are at home in *experimental* physics, where the reference to observers and circumstances of observation is legitimate and often explicit. Here we find two kinds of pragmatic interpretation: strict and adventitious. Let us start with the former. An expression such as "$f(x,y) = z$" can in principle be interpreted in this way: "The f-ness of x, as observed (or measured) by y, equals z." Since the formula makes room for the pragmatic referent y, the preceding interpretation is both strict and partly pragmatic. But, of course, any such interpretation must be only *partly* pragmatic if it is to count as a sentence in the language of physics, for this science happens to be concerned with physical systems. Secondly, the variable y must designate a possible observer, not a mythical one like the observer at infinity (or even worse, the continuous stream of observers) imagined by some field theorists. Thirdly, y must be a variable in the intuitive sense: that is, a change in the value of y must make some difference to the value of f. In short, subject to certain restrictions, *some* physical formulas can be given a *strict* interpretation which is *partly pragmatic*.

But the most common pragmatic interpretations found in the physical literature of our century are adventitious rather than strict: that is, they do not assign a meaning to every variable in a complex symbol but take the latter in block and match it with a pragmatic item from the outside. Thus given a sentence s belonging to a physical language, the following pragmatic interpretations of s are frequently met in the literature: (a) "s summarizes the measurements performed by a qualified observer"; (b) "Perform the operations necessary to test s"; and (c) "Act (analyze, measure, build, destroy...) in conformity with s." The reference to the physical object has been obliterated: everything points now to an active subject. Consequently the ideal of scientific objectivity seems to have been discarded.

While a strict pragmatic interpretation may contribute to deter-

---

[3] Note that "valid" does not amount to "right." Thus, an invalid or unjustified pragmatic interpretation may eventually prove to be right, for a theory may be built that justifies it. Conversely, a valid pragmatic interpretation may turn out to be wrong, for the theory supporting it may have to be given up.

mining the meaning of a symbol, an adventitious interpretation, whether pragmatic or not, fails to perform this function: it just prescribes or suggests, in a more or less precise fashion, a way of action. It does not tell us what the symbols stand for but what can be done with them—which is all technicians and Wittgensteinian philosophers care for. Secondly, for a sign s to be handled in an effective way and in conformity with one of its pragmatic interpretations, s need not make full sense to its prospective user even though it must make some sense to the person ultimately responsible for such use. Thus computations and even measurements can be executed with the help of computers that are given only pragmatic interpretations. But the programmer must be aware of the semantic interpretation of the symbols he handles, for otherwise he will be unable to write out any program and decode the machine's output. Thus if a sentence s expresses some property P of a physical system, any pragmatic interpretation of s for the use of, say, an automated measurement arrangement, requires not only an adequate semantic interpretation of s (i.e., one pointing to physical systems), but also its linking to a number of further sentences capable of expressing a way in which P can effectively be measured on a concrete physical system. (These additional sentences usually belong to theories other than the theory in which s inheres.) In other words, the design and execution of empirical operations, whether automated or not, involves the assignment of *pragmatic interpretations based on semantic interpretations*. In short, adventitious interpretations, even when legitimate, cannot replace semantic interpretations.

4. *Four Theses Concerning the Referent of Physical Theory.*

Before asking what the referent of a theory may be we must ask whether it does have a referent at all. There are two possible answers to this previous question: "Do physical theories have a referent?" One is in the affirmative, the other in the negative. The latter is, indeed, the conventionalist or instrumentalist view according to which physical theories are not about anything but are just data summarizing and processing tools, i.e., instruments enabling us to can information and grind out predictions. This answer is unsatisfactory for at least two reasons. First, it fails to tell us what kind of data physical theories are supposed to handle and what sort of predictions they are supposed to yield in contradistinction to, say, sociological theories. Second and consequently, a consistent conventionalist would not know how to go about testing a physical theory, for every such test presupposes knowing what the theory is supposed to be concerned with: indeed, if a theory intends to refer to, say, fluids, then it will be fluids rather than, say, atomic nuclei or wars which will have to be looked into. We shall therefore discard the conventionalist thesis.

A non-conventionalist should then have an answer to the question: "What are physical theories about?" Since the referent of a factual theory may be a physical object, or a subject, or some combination of the two, there are four possible and mutually exclusive answers to the question of the identity of the referent. These are:

(1) The *realist* thesis: a physical theory is about physical systems, i.e., it is concerned with entities and events that undoubtedly have an autonomous existence (*naive realism*) or else are assumed (perhaps wrongly in some cases) to have an autonomous existence (*critical realism*). In short, the physical interpretation of every nonformal formula in theoretical physics must be both *strict* (as opposed to adventitious) and *objective* (as opposed to subjective): every theoretical statement in physics is thus a *physical object statement*.

(2) The *subjectivist* thesis: a physical theory is about the sensations (*sensim*) or else about the ideas (*subjective idealism*) of some subject engaged in cognitive acts—in any case it is about mental states. In short, the physical interpretation of every formula in theoretical physics must be both *strict and subjective*: every theoretical statement in physics is thus a *mental object statement*.

(3) The *strict Copenhagen thesis*: a physical theory, or at any rate quantum theory, is about unanalyzable subject–object blocks. No absolute (subject-free, objective) distinction can be drawn between the two components of any such block: the boundary between them can be shifted at will. In short, the physical interpretation of every nonformal formula in theoretical physics, or at least in quantum theory, must be both *adventitious* (as opposed to strict) and *physico-mental* (as distinct from either physical or mental), for the observers and his conditions of observation must be read in every such formula even though the corresponding variables may be missing: every theoretical statement in physics is thus a *physico-mental* statement.

(4) The *dualist* thesis: a physical theory is about both physical objects and human actors: it concerns the transactions of humans with their environment (*pragmatism*) or the ways humans handle systems when intent on knowing them (*operationism*). In short, the physical interpretation of every formula in theoretical physics, whether *strict or adventitious*, must be *partly objective, partly pragmatic*: every statement in theoretical physics is thus a *partly physical and partly mental object statement*.

WHAT ARE PHYSICAL THEORIES ABOUT? 71

The realist thesis is the one that prevailed during the classical period of physics, and which was defended by Boltzmann, Planck, the later Einstein, and de Broglie. The subjectivist thesis was frequently defended by Mach (whose "elements" or atoms were sensations) and occasionally also by Eddington and Schrödinger. The Copenhagen thesis was advanced by N. Bohr and defended by his faithful followers—the qualifier being in order, for most of those who profess their loyalty to the Copenhagen school actually waver between theses (3) and (4) above. And the dualist thesis has been expounded in various ways by Peirce, Mach, Dingler, Dewey, Eddington, Bridgman, Dingle, and Bohr, as well as by hundreds of writers on relativity (who identify reference frames with observers) and quanta (who take apparatus for observers). The fourth thesis is, indeed, the nucleus of the official philosophy of physics, even though it has never been vindicated by a careful analysis of theoretical expressions.

The first three theses are *monistic* in the sense that each of them asserts that the reference class of a physical theory is metaphysically homogeneous (physical, mental, or physico-mental respectively) and moreover irreducible to entities of a different kind. The fourth thesis is *dualistic* in the sense that it postulates two mutually irreducible substances. Mathematically speaking, the reference class of a theory interpreted in a monistic spirit is homogeneous in the sense that all of the individuals or members of that set are assumed to be of the same broad kind: either physical, psychical, or psychophysical (see Sect. III, 1). On the other hand, the reference class of a theory cast in a dualistic spirit will be a cartesian product of at least two sets, at least one of which represents a kind of physical object while at least one other such set represents observers.[4] Thus e.g., the domain of an absolute probability function occurring in a physical theory will be interpreted in the following ways by the various semantical schools encountered in the philosophy of physics: the set of physical events of a kind (realism), the set of mental events of a kind (subjectivism), the set of irrational (unanalyzable, hence incomprehensible) phenomena of a kind (Copenhagen), and the set of pairs physical event–observer (dualism).

Every monistic thesis expresses a commitment to the hypothesis that there are entities of a certain kind: physical objects, minds, or psycho-physical complexes, as the case may be. The dualist, on the other hand, while claiming that physics is about both things and

[4] See, however, Sect. III, 3 for the impossibility of carrying out the dualist program.

actors, will refuse to acknowledge the independent existence of physical objects and will therefore come close to the Copenhagen doctrine. He will advance the methodological thesis that a statement concerning a thing in itself is untestable. And, basing himself on the verification doctrine of meaning that was fashionable four decades ago, he will conclude that such a statement is meaningless. The pragmatist, then, is neither a realist nor a subjectivist: he is an agnostic just like Kant. And, like the Copenhagen philosopher, the pragmatist maintains that a theoretical statement makes no sense unless it is accompanied by a description of the conditions of its empirical test. But, unlike the Copenhagen philosopher, the dualist distinguishes the subject from the object and he does not deride the attempt to analyze the subject–object interaction even though he may not care to perform it himself. Furthermore, the dualist is not prepared to acknowledge the existence of a third kind of stuff composed, in arbitrary varying proportions, of object and subject.

To discover which of the four preceding philosophical theses about the content of a physical theory is right, quotations from the Old Masters will not help, not only because arguments from authority are worthless but also because, as we saw before, every one of the four above theses enjoys the support of at least one great name. Worse: one and the same author may endorse two mutually incompatible theses in the same writing, without seeming to realize their difference. Thus Mach and Dewey wavered between subjectivism and dualism; and Bohr, who started out as a realist, became a subjectivist, oscillated later between dualism (as often represented by Heisenberg) and the strict Copenhagen thesis, and is said to have finally reverted to realism. Nor will general philosophical discussions be of much help, for the object of our inquiry is a special kind of human knowledge. Rather, we must take the bull by the horns and analyze physical theories and their components (concepts and statements). We shall proceed to sketch such an analysis with particular reference to quantum theory for, while the problem antedates this theory, it has become more acute with its birth.[5]

II. Identifying the Referent

1. *The Theoretical–Experimental Dichotomy.*

Although the four theses expounded in the last section concern the reference of a physical theory, all but the first, or realist thesis, are

---

[5] For detailed analyses of several major physical theories, see M. Bunge, *Foundations of Physics* (New York, Springer-Verlag, 1967).

meant to cover the whole of physics, both theoretical and experimental. Indeed, for the consistent subjectivist every self-contained physical formula is a mental object sentence; for the Copenhagen philosopher every such expression concerns an indissoluble mind–body compositum; and for the dualist every such sign is about both objects and subjects. On the other hand the realist claims that, while an empirical statement (e.g., an experimental datum) concerns both a physical object and an observer (or a team of observers), a theoretical statement fails to point to any subject whatever, the concern of theoretical physics being to account for the world such as it is, independently of being perceived or manipulated. In short, of all four only the realist makes a semantic difference between theoretical and experimental physics.

Moreover, the realist will probably point out that this distinction enables him to mark off the meaning of a formula from its test—two concepts that are hardly distinguishable for both the Copenhagen and the dualist philosophers. And he may add that this distinction between theoretical and experimental physics makes it possible to understand why theoreticians work only with their heads while experimentalists must, in addition, fumble around with pieces of equipment. Finally, the realist may also say that the very same distinction is necessary to understand why theories are not self-testing (as they should if subjectivism were true) and why any empirical test consists in contrasting and assaying theoretical predictions, on the one hand, and experimental results on the other. In any case, whether or not we end up by adopting realism, we should give it a chance to defend itself, accepting to make the theoretical–experimental distinction even if we intend to end up by denying that there is any.

Now, theories are certain sets of statements (infinite sets closed under deduction), and every physical statement contains at least one physical concept for otherwise it would not qualify as a physical statement. Therefore our semantic analysis of physical theories must begin with physical concepts. It may as well end with them, for this level of analysis is necessary and sufficient to disclose the referent of a physical theory: a theory is indeed about all and only those items referred to by the concepts with which the theory is built. A systematic investigation of this kind being out of the question in the present context, we shall restrict our attention to a few typical examples.

2. *The Referent of a Physical Quantity.*

The so-called physical quantities (or rather magnitudes) are said to "pertain" to a physical system of some kind or to be "associated" with such an object. This association of symbol with thing becomes obvious when it is necessary to name the components of a complex system, e.g., by assigning them numerals. Thus the *P*-ness of the *n*th component of a system may be designated by P_n.

These vague phrases may be elucidated by introducing two basic semantic concepts: those of referent and representation. Indeed, what is meant by saying that *P* "pertains to" or "is associated with" a physical system of a certain kind, is this: the concept (function) *P* represents a physical property (call it $\mathscr{P}$) of an arbitrary system σ of a certain kind, call it Σ. Hence Σ is the (intended) *reference class* of *P*. (In a moment we shall generalize this to the case of multiple referents.) The explicit mention of the referent is a reminder that, unlike functions in pure mathematics, those in theoretical physics may concern actual physical systems. And the explicit mention of the representation of a property $\mathscr{P}$ by a concept *P* calls our attention to the possibility that one and the same property (e.g., the electric charge) may be represented by alternative concepts in different theories. In short: *P represents* $\mathscr{P}$, which in turns *refers to* Σ. All this will presently be clarified and exemplified.

Let *P* be a function representing the property $\mathscr{P}$, or $P \triangleq \mathscr{P}$ for short. Now a function is not well-defined unless both its domain and its range are given. In the simplest case, of an invariant quantitative property, the function in question will be "defined" on a set of elements interpreted as so many physical systems, and into some set of numbers. The domain of this function will be either a set of individual systems (as in the case of the charge) or a set of pairs (as in the case of an interaction) or, in general, of *n*-tuples of physical systems. These sets are, of course, the reference classes of the concept *P*.

Example 1. In classical electrodynamics, the electric charge is represented by a function *Q* from the set Σ of material systems to the set R^+ of non-negative reals, i.e., $Q: \Sigma \to R^+$. Actually a further "independent variable" occurs in *Q*, namely the scale-cum-unit system *s*, which is usually specified by the context of the formula in which *Q* occurs. Hence, a correct analysis of the classical electric charge function is:

[1] $Q: \Sigma \times S \to R^+$

where *S* is the set of all conceivable scale-cum-unit systems. For example, for Σ = electrons and *s* = electrostatic units on a uniform metric scale,[6] one has the (usually unspoken) law statement:

---

[6] The concept of scale employed in the physical sciences differs from the one occurring in the philosophy of psychology after S. S. Stevens: the latter calls "scale" what the physicist calls "magnitude" or "quantity." For an explication of a physical scale, see M. Bunge, *Scientific Research* (New York, Springer-Verlag, 1967), vol. II, sect. 13.4.

WHAT ARE PHYSICAL THEORIES ABOUT? 75

[2] For every σ in Σ: $Q(\sigma, \text{e.s.u.}) = e \equiv 4.802.10^{-10}$

In any case, the reference class of Q is the set Σ of bodies.

Example 2. In all physical theories, the position (or else the position density) of a point of a physical system, whether it be a body or a field, a classical or a quantum-mechanical system, is represented by a vector valued function X of the elementary system, the reference frame, and time. In short, presupposing again a uniform metric scale,

[3] $X : \Sigma \times F \times T \to R^3$

where F is the set of physical frames and T the set of instants, i.e., R augmented with the elements $+\infty$ and $-\infty$. Since T is not a thing, the reference class of X is just $\Sigma \times F$. This is what is meant by the definite description "the position of σ relative to the frame f."

Example 3. The effective cross section of a particle of the kind A for elastic scattering by a particle of the kind B with wave momentum k relative to a frame f (e.g., the center of mass system), is a function σ (notice the change in notation) from the set of quadruples $\langle a, b, f, k \rangle$, where $a \in A$, $b \in B$, $f \in F$, and $k \in K$, to the positive real line. In short,

[3] $\sigma : A \times B \times F \times K \to R^+$

where K is the momentum range. For example, for $A = $ protons, $B = $ neutrons, and $f = $ center of mass frame, we have, to a first approximation, the well-known quantum-mechanical formula

[4] $\sigma_{PN}^{cm}(k) = 4\pi/k^2$.

Actually the CGS system has been presupposed. In any case, the reference class of σ is $A \times B \times F$.

Our semantic analysis so far favors a monistic interpretation: indeed, the reference class of a physical magnitude seems to be either a set of systems or a cartesian product of sets of systems. The observer is nowhere in sight. Moreover our analysis refutes the Copenhagen doctrine, insofar as no traces of the subject–object sealed unit have been detected. Only two views are upheld by our analysis: realism and subjectivism (see Sect. II, 2). The choice between the two cannot be made on the basis of our analysis alone, for the subjectivist will find no difficulty in identifying every system we call "physical" as a mental object.

Only an analysis of the sum-total of scientific activities tilts the balance in favor of realism.[7] Let the following reasons suffice here. (a) Every investigator starts out by admitting his ignorance of something he presumes, if only provisionally, to exist by itself and

---

[7] See M. Bunge, *Scientific Research*, ibid., Vol. I, Sect. 5.9, and Vol. II, Sect. 15.7.

waiting, as it were, to be discovered. (b) Every properly formulated theory starts by assuming that the reference class it is concerned with is nonempty, for otherwise the theory would be vacuously true. But this is no less than a (critical) commitment to the hypothesis that the theory has real referents. (c) No matter what subjectivist fancies the theoretician may indulge in when writing popular articles, the experimentalist is bound to take a realist attitude towards his experimental arrangements, the objects of his inquiries, and even his colleagues.

But, as we mentioned before, the observer concept, while absent from theoretical physics, does enter experimental physics. Thus, instead of the observer-free formula [2], we find in experimental physics statements like this: "The value (in e.s.u.) of the electron charge e, as measured by the experimental group g with the technique t and the instrument complex i, equals $(4.802\pm.001) \cdot 10^{-10}$." In short, instead of [2] we now have

[2'] For every σ in Σ examined by g: $Q'(\sigma, \text{e.s.u.}, g, t, i)$
$= (4.802\pm.001) \cdot 10^{-10}$,

where Q' designates the function whose values are measured values of the electric charge. Note that Q' differs from Q (the theoretical concept) not only numerically but structurally as well: it is a *different concept* altogether. In fact, while Q is a function on the set $\Sigma \times S$, Q' is "defined" on the set $\Sigma \times S \times G \times T \times I$, where G is the set of experimental groups, T the set of charge measurement techniques, and I the set of charge measurement equipments. Actually a sixth "independent variable" is involved in the experimental concept of electric charge, namely the sequence o of operations by which any given technique t is implemented by a group g with the help of the instrumental complex i. That o is not a phony or vacuous variable (in the sense of Sect. I, 2), is shown by the fact that changing its value usually does make a difference in the numerical result. Calling O the set of all such sequences of operations, we have finally, in contrast to [1],

[1'] $Q': \Sigma \times S \times G \times T \times I \times O \to R^+$

We now feel justified in drawing a general conclusion from the preceding analysis: *While the theoretical formulas are observer-free, the experimental formulas are observer-dependent.* More precisely: whereas any strict interpretation of a theoretical formula is *objective* (i.e., is cast in terms of physical concepts alone), experimental physics calls for a *pragmatic reinterpretation* of the same formula. Such a pragmatic interpretation, though adventitious (yet possibly valid) if bearing on a theoretical formula, is a strict interpretation in an experimental context, i.e., it is validated by substituting an experimental formula (such as [1']) for its theoretical partner (e.g., [1]).

Both the Copenhagen and the dualist interpretations of physical theories arise from a confusion between theoretical and experimental concepts, even though the latter rest on the former and are more complex than their theoretical bases. (In particular, a theoretical function may be regarded as the restriction of the corresponding measured value function to a certain set. Thus the Q in [1] is the restriction of the Q' in [1'] to $\Sigma \times S$.) This confusion may not be deplored by the Copenhagen philosopher, for whom everything is incurably irrational at bottom, but it defeats the very aim of the dualist, which is to avoid Platonizing. Indeed, if every theoretical formula is assigned a pragmatic interpretation, then it becomes impossible to contrast theory to experiment in order to test the former and design the latter. Moreover, since pragmatic interpretations are mostly adventitious, arbitrariness is apt to set in: everyone will feel entitled to read any formula as he pleases, independently of the structure of the formula: semantics will have no syntactic support. The realist and the subjectivist interpretations are free from these shortcomings. If we drop subjectivism for the reasons indicated a while ago, realism remains as the sole survivor. We should adopt then a *strict and objective* interpretation of every theoretical formula. Let us see how this approach fares in quantum theory, often alleged to be the grave of realism.[8]

3. *State Vector.*

It is generally agreed that the state vector or wave function ψ is a probability amplitude (i.e., that the square of its modulus is a probability density). Moreover, this, which is Max Born's "statistical" (actually probabilistic or stochastic) interpretation of ψ, can be proved from a certain set of postulates, whence it is far from being *ad hoc*,[9] hence avoidable if the standard formalism of quantum mechanics is kept. On the other hand there is no consensus about what ψ is a probability amplitude of. Sometimes ψ is regarded as concerning an individual system, others as referring to an actual or a potential statistical ensemble of similar systems, others as measuring our information or our degree of certainty concerning the state of an individual microsystem, or else of a microsystem–apparatus complex, or finally as summarizing a run of measurements on a set of identically

---

[8] For alternative realist approaches to quantum mechanics, see *Quantum Theory and Reality*, ed. by M. Bunge (New York, Springer-Verlag, 1967).

[9] M. Bunge, *Foundations of Physics, op. cit.*, pp. 252 and 262.

prepared microsystems.[10] In any case it is customary to assign ψ some such interpretation without caring whether it matches the structure of ψ and without making sure that the interpretation contributes to the consistency and the truth of the theory.

Yet it is possible to avoid the arbitrariness inherent in the adventitious interpretations of ψ and spot its genuine referents. The key is of course the hamiltonian operator H, since according to the central law of quantum mechanics (Schrödinger's equation or its operator equivalent), H is what "drives" ψ in the course of time. Now, H is not given from above but is what we want it to represent: a free electron, a carbon atom, a DNA molecule, or what not—and this in any hamiltonian theory, whether classical or quantal. We must then start by stating our claims or hypotheses concerning what it is that H, hence ψ, is going to be about. Some such claims will prove to be (approximately) true, others false: such is the life of theories.

Let us consider the mathematically simplest (but semantically most problematic) of cases: one-"particle" (or rather one-quanton) quantum mechanics. If the formalism turns out to be true under a certain interpretation of the basic (primitive) symbols, then the whole thing, formalism-cum-semantics, will be judged true of such individual systems even though its empirical test will call for the intervention of sentient beings manipulating (directly or by proxy) large collections of microsystems. In this theory H, hence also ψ, depend on time and on two sets of dynamical variables: those concerning the microsystem of interest (e.g., a silver atom) and those concerning the environment of the system (e.g., a magnetic field). If no such macrosystem is assumed to act on the given microsystem, i.e., if the latter is assumed to be free, then no macrovariables will occur in the hamiltonian (nor, consequently, in the state vector), no matter how much *ad hoc* talk of observers and measuring devices may be indulged in. Indeed, it is utterly arbitrary, hence a matter of blind belief, to claim that even though a given hamiltonian fails to contain macrovariables, actually it concerns an observing mind, or a mind-body compositum, or even a microsystem–apparatus complex. Any such interpretation, by failing to match the syntax of H and ψ, is adventitious: it involves vacuous or idle variables: it has a ghostly quality. Both the formalism of (elementary nonrelativistic) quantum mechanics and the set of its applications (e.g., to molecules) warrant only this analysis of every state vector:

[10] For a critical survey of a number of interpretations of the state vector, see M. Bunge, "Survey of the Interpretation of Quantum Mechanics," *American Journal of Physics*, vol. 24 (1956), p. 272, reprinted in *Metascientific Queries* (Springfield, Illinois; Charles C. Thomas, 1959).

[5] $\psi: \Sigma \times \Sigma' \times E^3 \times T \to C$

where Σ is the set of microsystems, Σ' the set of macrosystems, E^3 the configuration (Euclidean) space, T the range of the time function, and C the complex plane. Consequently the referent of this theory is $\Sigma \times \Sigma'$, i.e., the set of all pairs microsystem–macrosystem, with the proviso that the second coordinate of this couple may be empty, not so the first, for this would render the whole theory pointless. Every other interpretation is footless: it has no leg to stand on except the *dicta* of some famous scientist and their philosophical apologists.[11]

4. *Probability*.

Of all the adventitious interpretations of the state vector, the subjectivist or nearly-subjectivist one is the most resilient, so that it will be worth our while to consider it in some detail. A popular argument for the thesis that ψ must be subjective or at least partly so, is the following: "The state vector has only a probability meaning [*true*]. Now, probabilities concern only mental states: a probability value can only measure the strength of our belief or the accuracy of our information [*false*]. Hence the state vector concerns our minds rather than autonomous physical systems [*false*]." This argument is valid but its conclusion is false because its second premiss is wrong: indeed, a task of stochastic theories in physics is to compute *physical* probabilities (e.g., transition probabilities and scattering cross sections) and statistical properties (averages, mean scatters, etc.) of physical systems, not of mental events. In any case, it is surprising that the same scientists who have often or even consistently adopted a subjectivist interpretation of probability such as Bohr, Born, Heisenberg, and von Neumann have at the same time believed that they had overcome classical determinism, an essential ingredient of which is the thesis that probability is but a name for ignorance. It would seem that, ever since statistical physics and statistical biology were born, we have acknowledged randomness as an objective mode of becoming, formerly only of aggregates, now of individual entities as well. In any case, the subjective interpretation of probability has no place in physics and it presupposes classical determinism.[12]

---

[11] For a rich collection of authoritative quotations in support of the Copenhagen interpretation and a return to Bohr's ideas, see P. K. Feyerabend, "On a Recent Critique of Complementarity," *Philosophy of Science*, vol. 35 (1968), p. 309.

[12] Further classical remains responsible for much of the confusion and inconsistency in the usual formulations of quantum mechanics are those of particle and wave. See M. Bunge, "Analogy in Quantum Theory: From Insight to Nonsense," *British Journal for the Philosophy of Science*, vol. 18 (1967), p. 265.

Whether or not probabilities are to be interpreted as physical properties on a par with lengths and densities, is not a matter of opinion but one of mathematical and semantical analysis. Only an examination of the independent variable(s) of a probability function will tell us whether the function can be assigned a (strict) interpretation as a physical property, or as a state of mind, or as "pertaining" (referring) to some thing–mind complex. But although such an analysis will indicate the interpretation possibilities it will not suffice to find out whether any of them is an admissible interpretation. The latter will be the case only if the probability calculus, or rather the whole formalism including that calculus, becomes factually true under the given interpretation. And, again, this is not a matter of taste, or philosophical school, or arbitrary decision, but rather a matter to be settled by analysis and experiment.

Take, for instance, the expression "$Pr(x) = r$," where "Pr" stands for the probability function and r for a member of the real number interval $[0, 1]$. If x stands for a physical object, such as a state or a change of state, then "$Pr(x)$" will be a property of that object, and any reference to observers, their operations, or their states of mind, will be supernumerary. Only if x symbolizes a psychical event will "$Pr(x)$" stand for something mental. There is no place for two referents, e.g., a physical object and a psychical one, where there is a single independent variable. Absolute (unconditional) probabilities are then impregnable to strict pragmatic interpretation in terms of both physical objects and actors: in order to rope in a subject we need a further variable. This possibility is afforded by conditional probabilities.

The expression "$Pr(x|y) = r$," read as "the probability of x given y equals r," could be interpreted in either an objectivist, or a subjectivist, or finally a dualist way (namely as concerning a thing–subject couple). For example, if the context or, even better, the explicit interpretation rules and assumptions, indicate that x stands for a physical object (e.g. a state or an event) and y for an observer, then "$Pr(x|y)$" might be read as the probability that the physical event x happens provided the observer y is present, or given that the mental event y has occurred, or in some other dualist fashion. But, as remarked earlier, any such interpretation will be legitimate just in case (a) the theory contains both independent variables and specifies them, and (b) the theorems of conditional probability are satisfied under this interpretation, i.e., they are satisfactorily confirmed by observation. There is actually a third condition to be met as well, namely the one of relevance: one can always add an observer variable, but unless this variable does make a difference and its properties are specified by the theory, it will be a vacuous or phony variable. So much for the strict interpretation of probability.

A pragmatic interpretation is always possible, even for unconditional (or absolute) probabilities, and it is often necessary, but it is never strict, i.e., it does not "flow" from the formulas but must be superimposed on them in a way that overflows physical theory. What I mean is this. Surely some probability values must be checked by someone, either theoretically or empirically, or both ways, so that statements of the following kind ensue: "The probability value r for event x was checked by observer y with the means z." But this statement does not belong to the theory: it does not qualify as a strict interpretation of the formula "$Pr(x) = r$." Something similar holds for every physical property, not just for probability. Thus the physical statement: "The distance between the end-points x and y of the body z, as measured by the observer u with means v, equals $r \pm \epsilon$." In short, given a theoretical statement with a strict physical interpretation, it can be attached to any number of adventitious pragmatic interpretations. But none of the resulting pragmatic statements belongs to the theory, just as none of the parasites of a tree is a part of the tree. The popular operationist contention that only pragmatic statements are meaningful, so that meaning assignments require a reference to empirical operations, rests on a confusion between meaning and verification, a confusion that has long since been cleared up by philosophers.

5. *Interpretation and Estimation of Probabilities.*

It is widely held that the frequency interpretation of probability, i.e., the interpretation of probability values as relative frequencies, is what we want in science. But this is not quite so. Indeed, when reading probabilities in terms of relative frequencies we do not perform a strict interpretation but rather a valuation or (statistical) *estimation*. That is, we do not assert that probabilities *mean* frequencies but that they can (sometimes) be *measured* by frequencies. In this regard a probability does not differ from any other physical quantity: it is a construct whose numerical value is to be contrasted to a measured value. Moreover, as there is no unique measurement technique for any given physical magnitude, so there is no single way of estimating probabilities from statistical data: sometimes one counts frequencies, at other times one measures entropies, at some other times one measures spectral line intensities, sometimes one measures scattering cross sections, and so forth. The very theory in which the probability concept is embedded may (but usually does not) suggest ways of estimating probabilities. In most cases additional theories are needed to estimate probabilities from empirical data. But this is not peculiar to probability: it holds for other properties as well.[13]

There are five additional reasons for rejecting not only the frequency *theories* of probability (like von Mises' and Reichenbach's) which are mathematically untenable anyhow, but also the frequency *interpretation* of probability. First, what one seems to mean by "$Pr(x) = r$" in physics is something like the strength (measured by the number r) of the tendency or propensity for x to occur, quite apart from the number of times it is (actually or potentially) observed to happen. The latter count will serve the purpose of testing a probabilistic formula rather than the one of assigning it a meaning. Second, while probabilities can be properties of individuals (e.g., events), frequencies are collective properties, i.e., properties of statistical ensembles. Third, the formulas of probability theory are not satisfied exactly by frequencies, not even in the long run, which is always a finite run. (Remember that frequencies do not approach probabilities. Only the *probability* of any preassigned departure of a frequency from the corresponding probability decreases with increasing sample size. But this theorem holds only for a special kind of random process, namely a sequence of Bernoulli trials. Furthermore, the second order probability that the theorem is concerned with is not itself reducible to a frequency.) Fourth, probability and frequency *are not the same functions*, for whereas the former (if absolute or unconditional) is defined on a certain set E, the frequency is defined, for every sampling procedure s, on a finite subset E^* of E. (In short, $Pr: E \to [0,1]$, while $f: E^* \times S \to F$, where S is the set of sampling procedures and F the collection of fractions in the unit interval.) Hence it is not true that one gets a model or true interpretation of the probability calculus upon interpreting probability values as observed relative frequencies; at most we could say that we thus get a *quasimodel*. Fifth, if a stochastic theory (such as statistical mechanics, quantum mechanics, genetics, or some stochastic learning model) is construed as yielding frequencies, then there is no point in performing any measurements to check the theoretical formulas. (Likewise with all other physical concepts, for example the one of

---

[13] For some of the complexities of measurement and experiment, and in particular their dependence on theories, see M. Bunge, *Scientific Research, op. cit.*, vol. II, chs. 13 and 14, and "Theory Meets Experiment" in *The Uses of Philosophy*, ed. by M. Munitz and H. Kiefer (Albany, State University of New York Press, forthcoming).

eigenvalue of an operator representative of a physical property: if eigenvalues were interpreted as measured values, as the orthodox school has it, then there would be no point in carrying out any actual measurements.) What makes both theory and experiment indispensable is that they are radically different: that a theory is not a summary of experiments, and that no run of experiments replaces a theory. It takes the two of them to engender any new item of knowledge.

In short, neither the subjectivist nor the dualist interpretation of probability have a place in theoretical physics: what do have a place in it are the following strict and objectivist interpretations: the *propensity* interpretation (Popper[14]) and the *randomness* interpretation. On the former, a probability value is a measure of the strength of the tendency for something to happen: probability is just quantified potentiality, with reference to physical systems, whether simple or complex, free or under the action of other systems, and in particular whether under observation or not.[15] On the second interpretation, probability is the odds or weight of an event belonging to a random collection (e.g., a Markov sequence) of events. On either interpretation, the probability of an event is an objective property of it: it is inherent in things; likewise a probability distribution is interpreted as an objective (but potential rather than actual) property of a physical system. The difference between the propensity and the randomness interpretations is that the former is wider, for it does not require the events to be random, while the randomness interpretation holds only for random events and therefore calls for criteria allowing one to find out whether the given set of events is a random one. In other words, the randomness interpretation of probability may be regarded as the restriction of the propensity interpretation to the subset of random events. On either interpretation the probability of, say, a transition from one state of a system to another state is just as objective as the speed of the transition: it is not in any way linked to ignorance, or to uncertainty, or conversely to the strength of our beliefs (which are usually too strong anyway). We shall call the two interpretations by the name *physical probability*.

[14] Karl R. Popper, "The Propensity Interpretation of Probability," *British Journal for the Philosophy of Science*, vol. 10 (1959), p. 25 and "Quantum Mechanics Without 'The Observer'" in *Quantum Theory and Reality*, ed. by M. Bunge, *op. cit.*

[15] Actually this is my own version of the propensity interpretation, as found in *Foundations of Physics*, *op. cit.*, p. 90. Popper's version (*ibid.*) concerns the object-experimental arrangement compound and could therefore be mistaken as supporting Bohr's thesis of the inextricable unity of the two—as has in fact been interpreted by Feyerabend, *op. cit.* In a personal communication Sir Karl has indicated agreement with my reinterpretation.

Whether or not one is suspicious of the propensity concept, one must surely regard the probabilities occurring in physics as physical properties on a par with internal stress and electric field strength. The reason is this: all the independent variables of a probability function in a physical theory stand for physical systems or properties thereof. (Even time, the least tangible of all physical variables, can be elucidated in terms of events and frames.[16]) There is no way of smuggling the observer and his mind into a theoretical probability statement by arguing, for example, that quantum mechanics does not concern autonomous systems but rather a complex constituted by a microsystem, an experimental arrangement (which one, pray?), and the operator of the latter. First, because this is simply wrong: most quantum-mechanical formulas are about microsystems embedded in a purely physical medium (which is very often absent). This is not a matter for pronouncements *ex cathedra* but a matter of analysis of the formulas concerned, and this analysis will not be exhaustive unless the formulas are written out explicitly, i.e., in the axiomatic way that is so distasteful to the enemies of clarity.[17] A second reason is that even those formulas which do concern an object–environment complex (e.g., a molecule immersed in an electric field), fail to be about an observer proper, i.e., a psychophysical being. For, if they were, the quantum theory ought to enable us to predict not only the microsystem's behavior but the observer's conduct as well, which unfortunately it does not. In conclusion, there is no ground for asserting that the cognitive subject enters theoretical physics, in particular the quantum theory, via probability and the state vector. And if he does not use these doors, it is difficult to see how else he could get in.

III. Distinguishing Apparatus From Observer

1. *Approaches to Measurement Theory.*

Many authors describe a measurement as an interaction between an object and an observer, or even as a synthesis of the two. But whereas some writers mean by "observer" a cognitive subject with his full psychical equipment, others mean a classically describable

[16] M. Bunge, *Foundations of Physics, op. cit.*, Chap 2, sect. 3 and, with much more detail and accuracy, in "Physical Time: the Objective and Relational Theory," *Philosophy of Science*, vol. 35 (1968), p. 355.

[17] For axiomatizations of several physical theories, see n. 5. For the peculiarities and virtues of physical (as different from mathematical) axiomatics, see M. Bunge, "Physical Axiomatics," *Reviews of Modern Physics*, vol. 39 (1967), p. 463, and "The Structure and Content of a Physical Theory" in *Delaware Seminar in the Foundations of Physics*, ed. by M. Bunge (New York, Springer-Verlag, 1967).

apparatus, and still others prefer to keep silent, hence ambiguous. If a difference between an observer and his equipment is not made, and if an observer is allotted a supraphysical mind (e.g., an immortal soul), then measurement becomes a gate through which soul and spirit flow not only into the making of physics, but also into the things themselves, which thereby cease to be things in themselves. Indeed, a standard argument against realism is from the nature of microphysical measurement. We must therefore take a look at the theory of the latter, or better at the various programs for setting up a measurement theory, for there are several and none has been fulfilled. This we must do not only in the interest of epistemology but also in the interest of experimental physicists, for if they were indistinguishable from their equipments, then either they should be paid no salaries or they should be allotted no funds for the purchase and maintenance of experimental facilities.

Essentially the following approaches to measurement theory in relation to the quantum theory can be found in the literature.

(1) *Naive realism:* (a) basic measurements are direct, i.e., in no need of theories; (b) derivative or indirect measurements can be taken care of by the available physical theories supplemented with mathematical statistics; (c) upshot: no special measurement theories are required. *Criticism:* see the next point.

(2) *Critical realism:* (a) there are no direct precision measurements, particularly in microphysics; (b) any detailed theory of the measurement of a physical magnitude (e.g., time reckoning) or of the preparation of a physical system (e.g., a proton beam with a given velocity distribution) calls for a number of general theories as well as a definite model of the experimental equipment (e.g., a cyclotron theory is an application of classical electrodynamics or, if preferred, it is a piece of relativistic technology); (c) since measurements are specific and they involve macrophysical systems, genuine theories of measurement (unlike the phony ones found in some books on quantum mechanics) cannot help being specific and involving fragments of classical theories (e.g., classical mechanics and optics); (d) no adequate *general* theory of measurement is available, either in classical or in quantum physics and moreover it is doubtful that any can be developed, precisely because there are no general measurements and every macrophysical event crosses several boundaries between the various chapters of physics. This is, of course, a thesis of the present paper.[18]

(3) *Naive operationism* (textbook philosophy): (a) every physical

[18] See n. 13.

theory, in particular quantum mechanics, concerns actual or possible measurement operations and their outcomes; thus a hamiltonian operator represents an energy measurement and its eigenvalues are measurable energy values; (b) consequently there is no need for a special theory of measurement. *Criticism:* (i) there is both a structural and a scmantical difference between a theoretical magnitude and its experimental partner, if any (recall Sect. II, 2); (ii) if general theories did concern empirical observations, then one of the two—theories or observations—would be redundant and the choice of equipment should make no difference.

(4) *Radical operationism:*[19] (a) basic measurements are direct; (b) a basic theory, such as quantum mechanics, should be concerned with basic measurements and be derived from an analysis of the physics of measurement. *Criticism:* (i) there are no direct measurements, at least not on microsystems (see the above criticism of naive realism); (ii) scientific analyses, whether of concepts or of operations, far from being extrasystematic, are performed with the help of theories; (iii) in particular, an analysis of a measurement presupposes a number of theories, both substantive (e.g., electromagnetic theory) and methodological (particularly mathematical statistics).

(5) *Strict Copenhagen view:*[20] (a) a measurement process is one in which object, apparatus, and observer become fused into a solid block, so that they lose their identities; (b) this unity is peculiar to the quantum phenomenon, which is thus unanalyzable; (c) "the quantum-mechanical formalism permits well-defined applications referring only to such closed phenomena";[21] (d) a theory of measurement would attempt to analyze such a unity, distinguishing between subject and object and finding out the precise form of their interaction, thus destroying the irreducibility and irrationality that characterizes quantum phenomena; (e) consequently no attempt to build a quantum theory of measurement should be made.[22] *Criticism:* (i) although a measurement act does involve an observer (and a number of other things as well), physics is not about sentient beings but about physical systems, sometimes under control but most often free and in any case devoid of mental components; (ii) it would be

---

[19] Günther Ludwig, "An Axiomatic Foundation of Quantum Mechanics on a Non-Subjective Basis" in *Quantum Theory and Reality*, ed. by M. Bunge, *op. cit.*
[20] N. Bohr, *op. cit.*
[21] *Ibid.*, pp. 73 and *passim*.
[22] See the remarks of L. Rosenfeld—Bohr's successor in Copenhagen—in *Proceedings on Theory of Gravitation*, ed. by L. Infeld (Paris, Gauthier-Villars, 1964).

desirable to have a number of genuine detailed quantum theories of real (hence specific) measurement processes, theories capable of explaining and predicting the whole chain starting from an elementary event (e.g., a photochemical reaction) and ending up in an observable macroevent (e.g., the blackening of a photographic plate): to wish otherwise is sheer obscurantism.

(6) *Von Neumann's view:*[23] (a) a measurement process is an object-subject interaction characterized by the arbitrariness of the frontier between the two (i.e., a cut can be made for the purpose of analysis but its position is conventional); (b) rather than being an application of quantum mechanics and other physical theories, a quantum theory of measurement requires suspending the main postulate of the latter (Schrödinger's equation or its equivalent), adopting in its stead the projection postulate, according to which the measurement of an observable throws the state vector onto any of the eigenvectors of the observable concerned; (c) the ensuing theory of measurement is quite general and moreover it gives quantum mechanics its operational meaning. Since this view is supposed to be the standard one, we shall concentrate our attention on it.

2. The Standard Account of Measurement.

The usually accepted account of the measurement process is the one given by von Neumann in a book that passes almost universally, though wrongly, for offering an axiomatic and consistent formulation of quantum mechanics.[24] This seems to have been the first time the observer was systematically allotted a prominent role in the account of experimental arrangements. Von Neumann made it clear that by an observer he meant not just a measuring apparatus but a human subject capable of "subjective apperception."[25] He even thought it necessary to rope in the doctrine of psychophysical parallelism. Von Neumann also insisted[26] that the frontier or cut between observer and observed system can be displaced at will. More precisely, he proposed dividing the world into three parts: the observed thing I, the measuring apparatus II, and the observer III. The frontier, he claimed, may be traced either between I and the compound system II+III, or between the physical complex I+II and the psychophysical entity III. In either case (a) a measurement is regarded as

---

[23] John von Neumann, *Mathematische Grundlagen der Quantenmechanik* (Berlin, Springer-Verlag, 1932). English translation: *Mathematical Foundations of Quantum Mechanics* (Princeton, Princeton University Press, 1955).
[24] *Ibid.*
[25] *Ibid.*, p. 223.
[26] *Ibid.*, pp. 224 ff.

being something very different from, say, the action of an external magnetic field on a spinning microsystem—precisely because of the unpredictable, nay capricious, intervention of the conscious mind, and (b) the measurement process is neither controllable nor fully reducible to physics, for it involves subjective apperception and arbitrary choice.[27]

Inconsistently enough, this tripartite division of the world is not embodied into a theory: it is vacuous. In fact, *nowhere* in von Neumann's book are the properties of the observer (system III) specified, even sketchily: (a) his discussion of compound systems[28] which sets the stage for his treatment of the measurement process[29] concerns the "observed" object coupled to the measuring apparatus, i.e., I+II, a compound of physical systems with no admixture of mental components; (b) von Neumann says explicitly that the subject "remains outside the calculation."[30] Now, something that does not occur in the theory, yet is supposed to be its distinctive mark (as opposed to a classical measurement theory), is a phony item, a ghost, a hidden variable in the bad sense of this expression. But the cognitive subject is not the only ghost in von Neumann's theory, or rather pseudotheory, of measurement. An actual ingredient of it has a ghost-like quality as well: this is the state of the observed system before a measurement is actually performed. For, if this state is empirically unknown and moreover unknowable, then it should not occur in a theory that is supposed to abide by an empiricist philosophy. (On the other hand it can occur on any other philosophy, for it may be regarded as a hypothesis to be tested by observation.) Furthermore, to maintain, as von Neumann did, that a measurement brings about a transition from that unknown state into an unpredictable eigenvector of the measured "observable," is to explain the obscure by the more obscure. In any case, a sketch of a theory of highly idealized measurements of arbitrary magnitudes, surrounded by empty talk about idle observers, cannot pass for a theory of actual measurement even though it be approved of (but never used by) the bulk of the physical profession. Moral for the philosopher: Never go by what the scientist says he does.

A reason for the failure of von Neumann to provide a genuine theory of measurement is, of course, that there is no such thing as an arbitrary measurement. A second reason is that he accepted uncritically the orthodox interpretation of quantum mechanics he

[27] *Ibid.*, pp. 223 ff.
[28] *Ibid.*, Chap. VI, sect. 2.
[29] *Ibid.*, Chap. VI, sect. 3.
[30] *Ibid.*, pp. 224, 234.

learned from physicists, without realizing that this interpretation renders measurement theories redundant.[31] Indeed, according to that interpretation, an eigenvalue is not a value actually possessed by a system but rather a measured value.[32] Hence no separate theory of measurement should be necessary if the orthodox interpretation is adopted. Now, if eigenvalues are measured values, then eigenfunctions must represent states of systems under observation. On the other hand a general state vector (a linear combination of eigenfunctions or eigenvectors) must represent a state of a system before or after it is observed, particularly if the subjective interpretation of probability is embraced, as von Neumann did half of the time. He did not see that there is no point in building a whole theory (quantum mechanics minus measurement theory) centered around the equation of evolution of such unobservable states. Nor did he realize that the duality of his two kinds of processes, the one of collapse of the state vector upon measurement (process 1) and the other of smooth ("causal" in the barbarous standard terminology) evolution in accordance with the Schrödinger equation (process 2), contradicts the very philosophy he espoused, for one does not write out a whole theory concerning a process that is in principle unobservable. Finally, von Neumann did not see either that—as Margenau[33] pointed out long ago—all the actual calculations in quantum mechanics, in particular those which have been checked by experiment, concern processes of the second kind, namely those satisfying the Schrödinger equation, rather than processes of the first kind. Therefore, if a general quantum theory of measurement were possible, which is doubtful, the natural thing to do would be to drop von Neumann's projection postulate and apply Schrödinger's equation (or an equivalent) to the object–apparatus complex regarded as a purely physical two-system entity[34]—or, even better, to

---

[31] The reader should recall the strict Copenhagen view discussed in Sect. III, 1.

[32] We have argued in Sect. II, 2 that this interpretation is adventitious and invalid.

[33] Henry Margenau, "Quantum-Mechanical Description," *Physical Review*, vol. 49 (1936), p. 240. Actually what Margenau shows in this paper is that certain paradoxes disappear if the von Neumann projection postulate is dropped. That this postulate is never used, was pointed out by Margenau to the writer ten years ago in conversation.

[34] This is how the problem is treated in n. 5 and in the following papers: A. Daneri, A. Loinger and G. M. Prosperi, "Quantum Theory of Measurement and Ergodicity Conditions," *Nuclear Physics*, vol. 33 (1962), p. 297; D. Bohm and J. Bub, "A Proposed Solution of the Measurement Problem in Quantum Mechanics by a Hidden Variable Theory," *Reviews of Modern Physics*, vol. 38 (1966), p. 453; and H. J. Groenewold, *Foundations of Quantum Theory*, preprint of the Institute for Theoretical Physics, Groningen University.

treat it as a many-body problem to be handled by quantum statistical mechanics. In any case a measurement theory would be an application of a basic theory rather than a chapter of it. However, the very possibility of a general theory of measurement, whether classical or quantal, is problematic, because a universal meter would measure nothing in particular.

So we have this anomalous situation. First, it is claimed that only a discussion of empirical operations, such as measurements, can supply the content or physical meaning of the mathematical formalism of the quantum theory. This tallies with the obsolete verification doctrine of meaning but is inconsistent with the practice of designing, analyzing, and evaluating empirical operations in the light of theories. Second, the standard quantum-mechanical theory of measurement (von Neumann's) does not have the blessing of the proponents of the equally standard interpretation of quantum mechanics. Third, von Neumann's theory of measurement is practically nonexistent and it is supposed to contain a concept, the one of observer, that is extraphysical and moreover has not been incorporated into the (pseudo)theory: it remains outside the formulas of the latter, it hovers above them without getting actually mixed with the actual components of the theory. Fourth, no realistic cases have been handled with the help of von Neumann's theory of measurement. He himself gave a single example which, for being concerned with two mass-points, is not an example of a real measurement; he left the discussion of realistic, hence enormously more complicated, examples to the reader.[35] As a consequence this theory remains *untested*: indeed, it has failed to yield a single verified prediction.

In short, the standard quantum theory of measurement, which is alleged to enthrone the observer in theoretical physics, is a ghostly one altogether. Consequently the usual attempts to discuss the foundations of quantum mechanics, and in particular its meaning, in terms of the theory of measurement, are as ill-advised as the attempts to disclose the nature of man via theology. Worse, the point of measuring is to get down to particulars, which can only be done with the help of specific pieces of equipment. And any such particular measuring set-up calls for a specific theory. And any such specific theory is an application of a number of general theories: actually it is a set of general theories together with a definite model of the experimental situation. Hence no single theory can be expected to account for every possible measurement device, except in such a superficial way that it will be helpless to explain and predict the

[35] von Neumann, *op. cit.*, p. 237.

behavior of a single particular experimental arrangement. Therefore the strict Copenhagen view, according to which no time should be wasted in trying to build a quantum theory of measurement, is right though for the wrong reason. But no matter what stand one may take on this controversial issue, the philosophically important point is that no existing quantum theory of measurement[36] is concerned with The Observer, *pace* the repeated verbal attempts to smuggle Him into the picture.

3. *Experiment Presupposes Realism and Confirms It.*

Strange as it may seem, the opponents of realism try to argue from the most tangible aspects of physics, namely laboratory physics. The favorite arguments are these. "A physical quantity has no value unless it is measured; now, measurement is a human action; hence physical quantities acquire a precise value only as a result of certain human actions. Likewise, a thing is in no definite state unless it is prepared to be in a given state; now, a state preparation is a human action; hence physical systems adopt definite states only as a result of certain human actions."

These arguments, though popular, are circular, for their conclusions assert the same thing as their major premises. In fact, "to measure" and "to prepare" are pragmatic terms which the minor premises spell out. The major premises say all the non-realist wishes to assert, namely that whatever is, is so because somebody has decided to make it that way or, equivalently, that properties and states have no autonomous existence but are observer-dependent. Moreover these premises are false, for they rest on a confusion between being and knowing. Surely a magnitude has no *known* value unless it is measured. But this does not entail that it *has* no definite value as long as it is not being measured. The contrary thesis amounts to the claim that the researcher does not investigate the world but creates it as he proceeds, which is philosophically ludicrous, as it leads to subjective idealism and ultimately to solipsism.

The non-realist thesis is also mathematically untenable. In fact, when formulating a physical theory one will state, for example, that a certain property is represented by a real-valued function, and one will assume or hope that measurements will be able to sample such values at least within an interval of the whole range of the function. One assumes, in other words, that the function *has* certain values all the

---

[36] This holds also for the theory formulated by F. London and E. Bauer in *La théorie de l'observation en mécanique quantique* (Paris, Hermann, 1939), even though the authors follow von Neumann's practice of toying with the observer concept and even though they throw in Husserl's idealism for good measure.

time for, if it did not have them, it would not be a function—by definition of "function." Similarly with the operators assumed to represent dynamical variables: they are supposed to have definite eigenvalues even while no measurement of such properties is being performed, for otherwise they would not be well-defined mathematical objects. This does not entail that a physical system has at all times a sharp position and a sharp velocity (or, in general, that it is at every instant in a simultaneous eigenstate of all its "observables"), only we do not happen to know such precise values. Since in quantum mechanics the dynamical variables are random variables, they have definite distributions (even for a single physical system) rather than definite numerical values. But these distributions and, in general, the bilinear forms built with the operators and the state vectors, do have definite values at each point in space and time, for they are ordinary point functions.

In short, the thesis that the values of functions and the eigenvalues of operators are measured values is mathematically untenable. Surely the decision to measure or to prepare a system, as well as the ensuing laboratory operations, are the doings of humans, and the outcomes of these actions will depend on them as much as the outcome of any other human action. But humans are part of nature, their action on their environment is efficient only insofar as it is based on some knowledge of nature, and only the physical aspect of such actions is relevant to physics: minds have no direct action on things and, even if they had, physics would not be competent to account for them. Surely the act of preparation modifies the initial state of the thing, whether or not it is a microsystem; but for such a change to occur the thing must be available or it must be produced out of other things that were there to begin with; also, the change must be a thoroughly real one even when steered by a subject.

With the exception of extreme subjectivists, who hope to get away without any empirical operations, everyone agrees that measurement and experiment are essential to physical research. Now, in order for any such operation to furnish genuine empirical evidence, it must be real: dreams and gedankenexperiments can be heuristically valuable but they prove and disprove nothing. In other words, the least one will do when assessing an experiment is to ascertain whether the experimental set-up is actual, otherwise one will speak of a plan for an experiment or even of a fraud. Of course, any experimental arrangement is artificial in the sense that it is planned, made, and controlled by humans, either directly or indirectly. But so is a car and so is an artificial satellite, and yet nobody would mistake such artifacts for observers. Now, we cannot satisfy ourselves that a certain experimental set-up is real unless its immediate environment is actual as well, for otherwise there would be no point in constructing insulators and in making temperature and pressure corrections, in inspecting the system for external disturbances and leakages, etc. Furthermore, every component of the system must be real for the whole to be real. If the components of a complex system were mental rather than physical, they would give rise to a psychical whole. This contradicts the claim of the Copenhagen philosophers that, while macrosystems (e.g., apparatus) may be real, their atomic constituents lack autonomous existence. Of course one often makes the mistake of believing that something is out there while it is actually missing. But errors of this kind can eventually be recognized as such, and such corrections show how much store we do set by the assumption that in the laboratory one handles real things.

In short, experimental physics assumes the reality of the objects it manipulates, and it tests some of the theoretical hypotheses made about the existence of physical systems. Experimental physics has no use for a physical theory that makes no existence assumptions, and theoretical physics can expect no help from experimentalists who are not willing to soil their hands with real things.

IV. Four Possible Styles of Theorizing

1. *The Realist and the Subjectivist Versions.*

In order to better assess the merits and demerits of the various philosophies discussed so far, we shall try to formulate in a cogent way (i.e., axiomatically) a very simple theory in four different guises, each corresponding to one of those philosophies. (This will have the side effect of underpinning the thesis that scientific research is far from being philosophically neutral.) We shall start with the realist and the subjectivist theories, which can be dealt with jointly because of their unambiguous monistic character.

Let the theory concern a physical system (alternatively, a subject) which either is in one of two states named A and B, or jumps from one of them to the other in such a way that each of the four possible events, $\langle A, A \rangle$, $\langle A, B \rangle$, $\langle B, A \rangle$, and $\langle B, B \rangle$, has a definite probability. (The first and the fourth are, of course, null events.) Five specific primitive (undefined) concepts will do the job: the set Σ of systems (alternatively, of subjects), a state function S, two constants A and B, and the probability function Pr. The difference between the two theories, the realist and the subjectivist one, lies in the referent: in the former case the reference class Σ is interpreted as a set of physical systems, while in the subjectivist case it is interpreted as a set of subjects. Accordingly the functions S and Pr become either properties of a physical system or properties of a subject. To save space the subjectivist interpretation will be indicated in parentheses and in *italics*. Only the axiomatic foundations will be laid down.

Axiom 1. There are physical systems (*subjects*) of the kind Σ. [In a slightly more detailed way: (a) $\Sigma \neq \emptyset$. (b) Every $\sigma \in \Sigma$ is a physical system (*subject*).]

Axiom 2. Any physical system (*subject*) of the kind Σ is in either of two *states* (*states of mind*): A and B. [More explicitly: (a) S is a many-to-one function from Σ into $\{A, B\}$. (b) A and B represent states (*states of mind*) of a physical system (*subject*) of the kind Σ.]

Axiom 3. (a) Pr is a probability measure on $\{A, B\}^2$. (b) The probability of any pair in $\{A, B\}^2$ is nonvanishing [all transitions are possible]. (c) $Pr(\langle A, A \rangle) + Pr(\langle B, B \rangle) = 1$. (d) $Pr(\langle A, B \rangle)$ represents the strength of the tendency or propensity (*observed relative frequency*) with which a physical system (*subject*) in state (*state of mind*) A jumps into state (*state of mind*) B, and similarly for the other probability values.

The ostensive differences between the two theories are these. (a) While the realist theory concerns an idealized physical system (a model of plenty of real situations), the subjectivist theory concerns an idealized subject (hardly a suitable model of anyone save an extreme moron). (b) While the realist theory informs about physical events, the subjectivist one informs about psychical events. (c) While the realist theory involves transition probabilities that can be checked by observing frequencies of external events, the subjectivist theory involves introspectively observable transition frequencies. (d) While the realist theory is testable in a physical laboratory, the subjectivist one is not testable in this way.

Both theories are phenomenological or black box theories in the sense that neither accounts for the transition mechanism. But they can be deepened so as to explain the transitions. In either case such a deepening calls for the introduction of new basic concepts and correspondingly of new postulates. (Remember the unspoken rule: For every new primitive, at least one new formal and one new semantical postulate.) Thus the realist theory can be expanded into a stronger theory explaining the transition probabilities in terms of, say, the state occupation number. For example, the probability of the event $\langle A, B \rangle$ could be set proportional to the occupation number of the state A and inversely proportional to the occupation number of B.

Or else a hidden variables theory might be set up: a theory containing further variables and equations of evolution for them that would explain both the existence of the states and the transitions between them. Any such stronger theory would still be a physical theory. On the other hand the subjectivist theory might be expanded in either of the following opposite directions: the new variables could be further psychological concepts, or some of them could be neurological (physiological) concepts. In the first case a homogeneous extension would be obtained: the new theory would remain within psychology. But in the second case the stronger and deeper theory would have a mixed character: it would contain both psychological and physical (or rather neurophysiological) variables, so that it would describe a two-level system. A still further extension might be able to analyze every psychological variable remaining in the former extension, in neurophysiological terms. Let us risk the following conclusions: any deepening of a realist theory retains its physical character, while some attempts to deepen a subjectivist theory change its character, thus defeating the philosophy of subjectivism. In other words, it would seem that subjectivism can be kept at the price of avoiding further deepening, which is not the case of realism. But we are not now concerned with depth.[37] Our aim was just to show that a theory can be cast either in realist or in subjectivist terms. We shall presently see that none of the two other philosophies we have been discussing allows this.

2. *The Copenhagen Predicament.*

In a theory built in the pure Copenhagen style there should be a single reference class: the set of sealed units constituted by the object, the observation set-up, and the observer. At first sight there should be no difficulty in obtaining the Copenhagen version of any given physical theory, such as the one expounded in the last section: it would seem that a reinterpretation of Σ as the set of trinities should suffice. As a matter of fact there are two technical obstacles in the way, one of a formal nature, the other semantical.

The mathematical obstacle to the Copenhagenization of theories is this. The claim that the referent of a theory is single (equivalently, that its reference class is homogeneous in the sense of Sect. I, 1), and furthermore unanalyzable, contradicts the claim that every "quantity" (magnitude) is relational in the sense that it concerns not just the

[37] For a preliminary explication of the depth concept see M. Bunge, "The Maturation of Science" in *Problems in the Philosophy of Science*, ed. by I. Lakatos and A. Musgrave (Amsterdam, North-Holland, 1968).

physical system of interest (e.g., an atom) but also its (artificial) environment and the observer in charge of the latter. These two claims of the Copenhagen school are obviously mutually contradictory, for the first boils down to the assertion that the domain of the functions (e.g., probability distributions) concerned involves a homogeneous set of indivisible blocks, while the second claim boils down to the assertion that that domain involves the cartesian product of the set of physical systems by the set of apparatus and the set of observers. So much for the mathematical difficulty.

The refusal to analyze the referent *unum et trinum* renders the interpretation task hopeless, for the properties to be assigned to that referent are neither here nor there: they are neither strictly physical nor strictly psychological. This is why the Copenhagen doctrine is as obscure as the doctrine of the trinity, according to which the Father (Apparatus), the Son (Microsystem), and the Holy Ghost (Observer) are united in one Godhead (Quantum Phenomenon). Take, for instance, the notion of state occurring in the microtheory expounded in the previous subsection. While in the realist (alternatively the subjectivist) interpretation A and B stand for physical states (alternatively mental states), of a system of a definite kind (physical or psychical), in the Copenhagen interpretation they should represent total or psychophysical states of the block: system-apparatus-observer. But no existing science accounts for such complex (yet unitary) entities.

In conclusion, it is impossible to build a *consistent* theory in the Copenhagen style. In other words, the Copenhagen interpretation of the quantum theory is inconsistent,[38] and moreover incurably so. Fortunately the baby—quantum mechanics—need not be thrown away together with the bath water.[39]

3. *The Dualist Version*.

Let us return to the microtheory approach discussed in Sect. IV, 1. Its axiomatic reformulation in a dualist (e.g., operationist) spirit would require two further distinct sets: the set I of instruments and the set O of observers or operators. These various items should be regarded as interacting but also as distinct. (If they were indistin-

---

[38] For some of the inconsistencies of the Copenhagen interpretation of quantum mechanics, see ns. 5, 10, and 12 as well as "Quanta and Philosophy," *Proceedings of the 7th Inter-American Congress of Philosophy*, Vol. I (Québec, Presses de l'Université Laval, 1967).

[39] See the author's realist axiomatization of quantum mechanics in his *Foundations of Physics, op. cit.*, ch. 5, and in his *Quantum Theory and Reality, op. cit.* Prof. Erhard Scheibe of Göttingen has undertaken to improve this axiom system.

guishable, if they constituted a solid block, they could hardly interact.) Hence the corresponding concepts must be taken as mutually independent primitives. The dualist version of our microtheory would then be based on seven rather than five undefined concepts.

Now for an axiom system to be satisfactory, it must contain axioms specifying both the mathematical structure and the factual meaning of every one of its basic technical terms. (This may be called the condition of primitive completeness.[40]) This is well-nigh impracticable in the case of the additional primitives I and O, and even if it were feasible it would be hardly desirable. It is impracticable because, while Σ is handled by a strictly physical theory and moreover a well-defined one, I and O require going far beyond that theory. In fact, the characterization of any apparatus in theoretical terms calls for a whole assembly of fragments of different theories. Likewise, the specification of any observer would take all the sciences of man: anthropology, psychology, sociology, etc. The theory would then acquire a gigantic size in case it could be developed at all. The dualist program is therefore infeasible. It is not desirable either, for the following reasons. First, it would render general theories impossible, for a general theory is one that is not tied to any special kind of experimental set-ups. Second, the dualist program would render the progress of physics dependent on the state of the sciences of man—whence, if adhered to at the end of the Renaissance, physics would not have taken off. After all, modern physical science was born in opposition to anthropocentrism.

In conclusion, of the four conceivable types of theorizing two are impracticable: the Copenhagen and the dualist ones. The realist and the subjectivist approaches are feasible but only the former yields objective, testable, and in principle improvable theories.

V. Conclusion: Realism Upheld

We started by distinguishing two kinds of interpretation of physical symbols: strict interpretation which matches the mathematical structure of the corresponding idea, and adventitious interpretation which overflows it. We showed that, while in theoretical physics only strict interpretations are warranted, adventitious interpretations (e.g., in terms of operations) are called for in experimental physics, but they are valid only insofar as they are backed by theories (e.g., theories accounting for the operations).

[40] M. Bunge, "Physical Axiomatics," *op. cit.*

We then applied the previous distinction to some basic physical concepts. The upshot was that the only strict interpretations in theoretical physics are either realist or subjectivist, all others being adventitious. But we showed that there is no ground for the subjectivist interpretation of two functions that pass for being the gates through which the mind enters the physical picture, namely the state vector and probability. This we did by examining the independent variables, i.e., the domains of those functions, as well as by recalling some of the presuppositions and goals of scientific research. The discarding of subjectivism left us with realism as the sole viable philosophy of physics.

Next we explored the possibility of casting one and the same theory in each of the four competing philosophical molds: realism, subjectivism, the Copenhagen view, and dualism (in particular operationism). It turned out that, while the first two projects are viable, the subjectivist one does not lend itself as easily to generalization and deepening, and in any case is incurably untestable, hence nonscientific. As to the Copenhagen version, it proved impossible without contradiction, and the dualist (in particular operationist) formulation proved impracticable. Once again realism was vindicated as the sole realistic philosophy of physics.

Finally we turned our attention to measurement theory, often said to be another door through which the spirit enters our new picture of the world. We found that the standard theory (von Neumann's) is ghostly on more than one count: it hardly exists as a realistic theory of actual measurements and it talks about an observer that is supernumerary, as it occurs nowhere in the formulas. Here again, our analysis has upheld realism and, in particular, the trite yet important thesis that physics is about physical systems—notwithstanding the nonrealist phraseology that so often surrounds physical formulas and physical operations.

Now, there are a number of realist views (hardly theories) of knowledge. Which one does our semantical and methodological analysis support? The answer is, of course, *critical realism*. This view is characterized by the following theses:

(1) There are things in themselves, i.e., objects the existence of which does not depend on our mind. (Note that the quantifier is existential, not universal: artifacts obviously depend on minds.)

(2) Things in themselves are knowable, though partially and by successive approximations rather than exhaustively and in one stroke.

(3) Knowledge of a thing in itself is attained jointly by theory and experiment, none of which can pronounce final verdicts on anything.

(4) This knowledge (factual knowledge) is hypothetical rather than apodictic, hence it is corrigible and not final: while the philosophical hypotheses that there are things out there, and that they can be known, constitute presuppositions of scientific research, any scientific hypothesis about the existence of a special kind of object, its properties, or laws, is corrigible.

(5) Knowledge of a thing in itself, far from being direct and pictorial, is roundabout and symbolic.

These theses are all that critical realism *lato sensu* commits itself to and all that our analysis supports. Beyond this there is ample room to work out genuine theories (hypothetico-deductive systems) of knowledge preserving and spelling out the preceding theses. Critical realism thus keeps the 17th century distinction, exploited by Kant, between the thing in itself (such as it exists) and the thing for us (such as it is known to us), but drops Kant's theses that the former is unknowable and that the thing for us is identical with the phenomenal object, i.e., with appearance. Indeed, critical realism maintains: (a) that the thing in itself can be known in a gradual fashion, and (b) that the thing for us is not the one presented to the senses but the one characterized by scientific theory. Moreover, critical realism does not assume that the thing in itself is knowable as such, i.e., without our introducing any distortion (removal and/or addition of traits). What distinguishes critical realism from other varieties of realism is precisely the recognition that such a distortion is unavoidable, for ideas are not found ready made: we think them out laboriously and correct them endlessly or even give them up altogether. There would be no point in going through this process if we were able to grasp physical objects (e.g., electrons and galaxies) exactly as they are—even less if we were able to create them in the act of thinking them up. And there should be little hope concerning future science if, as the Copenhagen philosophy maintains, we have already reached the final frontier—the unanalyzable, irrational quantum phenomenon. Critical realism encourages us to look beyond every theory, however successful and therefore perfect it may look at any given time. In particular, it encourages the exploration of new pathways in fundamental physics — which, every one seems to agree, can use some radically new ideas.[41]

McGill University

[41] Written for the conference on Quantum Theory and Beyond held at the University of Cambridge, England, July 1968, and read at the conference on Foundations of Quantum Mechanics held at the New Mexico Institute of Mining and Technology, Socorro, New Mexico, August 1968.

V

The Plausibility of the Entrenchment Concept

BERNARD R. GRUNSTRA

GIVEN the definition of entrenchment employed in Nelson Goodman's prospects for a theory of projection (pp. 87, 94), it is clear that the fear of some is ill-founded that "grue" may come to be much better entrenched than "green" through the literature about it, with resultant disastrous consequences for the theory.[1] Accordingly this paper will fearlessly bring the topic up again. The principal purpose is to inquire how the entrenchment proposal could be considered a reasonable component of a solution to the problem of justifying projection; i.e., to ask how and in what sense predicate entrenchment could plausibly be considered an indispensable and normative factor in the assessment of hypothesis projectibility. As an auxiliary effort the paper will begin with a sketchy and inconclusive comparison of the projection problem raised by the "grue" example with the problem of alternative hypotheses generally, in particular with the curve-plotter's problem. The comparison suggests that perhaps not all hypothesis projection preferences can be correlated with single-predicate entrenchment superiority, but that in any case the study of the curve-plotter's problem is, if promising, also problematic with respect to understanding the role of entrenchment in projection. Since the entrenchment solution of the quest for projection justification was a response to the point of the "grue" example, we next study the features of that example with a view to making the statement of its point more plausible than many misunderstandings have permitted. Next we consider other proposed solutions to the projection problem, some which do accept the point of the "grue" example and some which do not. In the light of their weaknesses the entrenchment proposal gains strength by the comparison. Finally we attempt to allay discontent with the entrenchment solution itself, first by noting the "anticipatory" significance of our

[1] Except where the context indicates otherwise, numbers appearing alone between parentheses are page citations of the second edition of *Fact, Fiction, and Forecast* [4]. Numbers in brackets identify references.

"descriptive" predicates, secondly by assimilating the proposed entrenchment normativeness to that of a methodological principle in common use in everyday life and the sciences.

I. The Relevance of the Curve-Plotter's Problem

1. The cutting edge of Goodman's critique of prior confirmation theory appears in his apparent demonstration that any projection whatsoever can be justified on the basis of any evidence whatsoever if we have only the existing logico-syntactical induction rules or confirmation criteria to help us (pp. 74, 75). Both the strength of his attack and its far-reaching implications for inductive theory have apparently often been missed by commentators because of their preoccupation with the idiosyncrasies of his illustrative examples. These examples employ unfamiliar predicates of an abnormal breed, such as the now famous pair, "grue" and "bleen." Furthermore there is a kind of informality about both the specification of these predicates and the comparison of their effect on the confirmation criteria with that of normal predicates which amounts to an underspecification of the illustration in matters of detail. Commentators who have in one way or another supplied a fuller specification have frequently been misled either to believe that inductive theory had not been successfully challenged, or else to propose inadequate remedies. Many of the consequent misapprehensions have appeared in the literature and many of these have been answered by Goodman himself or by still other commentators. I shall not attempt to review them here.[2]

What is of more interest here is the possibility that Goodman's case, or one that seems much like it, can be made quite tellingly with the use of predicates that do not seem particularly abnormal and do not allow the glib dismissal of his argument as depending on predicates that no one, particularly practicing scientists, would consider seriously in inductive practice anyhow. This possibility is suggested by similarities with the curve-plotter's problem, most recently remarked by Skyrms ([13], pp. 52–69) and by Hullett and Schwartz [10].

If we consider Goodman's charge apart from his examples, it seems to come to this: Since different evidence statements can describe the same empirical data, we can, via criteria governing evidence statements, use this data to support widely different (and incompatible) projections. But it is a familiar feature of data-plotting

[2] How strong Goodman's case remains under appropriate fuller specification of detail can be seen in a clear article by Leblanc [11].

situations that through a given set of data points an indefinite number of distinct curves may be drawn, each of which leads to different predictions for unexamined points. Is not the essential thrust of the "grue" example, therefore, somehow the same as that made long ago by examples of alternative hypotheses? If so, our study of the role of predicate entrenchment in the assessment of hypothesis projectibility could perhaps be furthered with the use of more familiar, realistic, and acceptable examples than the "grue" example amounts to.

This advantage may not lightly be claimed, however. It depends on establishing how the "grue" problem relates to the alternative hypothesis problem generally, and how the entrenchment solution relates to this general problem. There is reason to think that establishing what these two relationships are calls for considerable discussion. It is not my purpose here to provide such a discussion, but I do want to say enough to indicate why it seems needed.

2. First there is the fact that for a projection problem drawn in terms of competing predicates one can provide a correlated projection problem in which the competition seems to lie elsewhere. Goodman introduced the projection problem in terms of competing predicates because to do so showed the deficiency of existing confirmation criteria even when strengthened by a suggestion of his own (p. 71). But to project (1) "All emeralds are grue" is, on the assumption that nothing is both blue and green, equivalent to projecting (2) "All emeralds examined before t are green and all emeralds not examined before t are blue." On Hempel's Satisfaction Criterion [9] the latter hypothesis is as well supported by evidence statements about all the emeralds examined before t and found green as is the hypothesis (3) "All emeralds examined before t are green and all emeralds not examined before t are green," which is equivalent to (4) "All emeralds are green." How is (2) to be eliminated in favor of (3)?

If we appeal to the equivalence of (1) and (2), and of (3) and (4), and the superior entrenchment-based projectibility of (4) relative to (1), then we provoke the following questions. (A) How do we know equivalences are alright to use, since by Goodman's own modification of Hempel's Satisfaction Criterion, equivalent hypotheses are not to be expected to be equally well-confirmed by the same evidence? (p. 71).[3] (B) If equivalences are all right to use, why can't we use the

---

[3] See also Scheffler [12], pp. 286–291, and Hanen [8]. Hypothesis (2) must be distinguished from "All emeralds are examined before t and green or not examined before t and blue." The latter is disfavored in competition with (4) on the same grounds as (1), namely the inferior entrenchment of the extension of the consequent predicate (pp. 95, 97).

THE PLAUSIBILITY OF THE ENTRENCHMENT CONCEPT 103

equivalences in the opposite direction, justifying the selection of (4) over (1) by the selection of (3) over (2)—and then try to establish the warrant for the latter selection in some non-entrenchment-based way?

If instead we attempt to establish the superiority of (3) over (2) by some non-entrenchment-based criterion, i.e., by a syntactic criterion obtained by some suitable augmentation of the Satisfaction Criterion, these questions arise. (C) Assuming we try to limit ourselves to the Selective Confirmation Criterion ([12], pp. 286–291), what is the proper form of a contrary of (2) which will help us to eliminate (2) as not selectively confirmed, but to retain (3) as selectively confirmed—supposing that is desirable? (D) On the same assumption, whether or not we can retain the selective confirmation of (3), what equivalences are we allowed to use, since on the Selective Confirmation Criterion not all will do? (E) Why is it that a difference should be drawn by such diverse methods between hypothesis pairs as equivalent as (4) and (1), or (3) and (2) are in many ways? The possibility of the questions just raised suggests that it may be no shortcut in understanding entrenchment to appeal to the alternative hypothesis problem in its general form.

3. A further consideration of the curve-plotter's problem suggests the same thing. In the form in which Hullett and Schwartz [10] cite this problem it becomes clearer that the point of the "grue" example is independent of the use of "time t" in its characterization.[4] But it remains the case that one is still appealing to abnormal predicates (unless one has switched to a non-predicate-competition example, such as that just discussed). Suppose we have examined nine samples of gas at constant pressure P_0 and constant temperature T_0, nine at higher pressure P_1 and temperature T_0, ... and nine at still higher pressure P_{10} and temperature T_0. After allowing for errors we agree that each of the nine investigated at a given pressure have the same volume, and that where P represents any of the indicated pressures and V the volume found to correspond to it, V is discovered to vary inversely with the pressure according to the rule (i) $PV = k_0 T_0$, where k_0 is a fixed constant. But then the 99 examined samples also obey the following rules (among many others), none of which is projectible:

(ii) $(P \leqslant P_{10} \ \& \ PV = k_0 T_0) \lor (P > P_{10} \ \& \ PV = 2k_0 T_0)$

---

[4] This was clear already from these other predicates Goodman uses: "in this room and English-speaking" (p. 37), "in this room and a third son" (p. 37), "in zig A" (applies to things in some "helter-skelter selection" of marbles, p. 104), and "bagleet" (applies to naval fleets and to a particular bagful of marbles, p. 111).

(iii) $((P=P_0 \lor P=P_1 \lor \cdots \lor P=P_{10})$ & $PV=k_0T_0) \lor$
$(\sim(P=P_0 \lor P=P_1 \cdots \lor P=P_{10})$ & $PV=2k_0T_0)$

(iv) (x is any of the 99 examined (k_0, T_0) samples, suitably identified & $P(x)V(x)=k_0(x)T_0(x)) \lor$ (x is a (k_0, T_0) sample but not one of the 99 examined samples & $P(x)V(x)=2k_0(x)T_0(x)$).

However, Goodman has made clear that the trouble he was pointing to was not to be attributed to unfamiliar predicates but to unentrenched extensions (pp. 95, 97). "Examined before t and green or not so examined and blue" is just as bad as "grue," and "examined before t and grue or not so examined and bleen (or, green)" is as good as "green." The predicate "$(P \leqslant P_{10}$ & $V=k_0T_0/P) \lor (P > P_{10}$ & $V=2k_0T_0/P)$," especially in competition with "$V=k_0T_0/P$," is as abnormal as "grue."[5]

4. But then one wonders if one can make the point of the "grue" example in the context of the curve-plotter's problem without using abnormal predicates. While it may not be sufficient to the creation of a predicate that will make the point "grue" is used to make that the predicate's extension be poorly entrenched, it is perhaps necessary. It is not clear, therefore, that we can study the role of predicate entrenchment in projectibility assessments with the help of a curve-plotting problem using competing normal predicates of reasonable entrenchment.

This seems borne out if we investigate such a problem. Consider a set of five data points, taken for some measurable property R at different values of some measurable property Q, such that when R values are plotted against Q values, the points appear to fall at irregular intervals along a straight line parallel to the Q-axis. One is inclined to say that R is constant with Q, and to proceed to draw a straight line through all five points parallel to the Q-axis. There are however, an indefinite number of other graphing possibilities. Let curve A be a sine wave of some frequency f such that it goes through zero displacement from its average value just at the data points; let curve B be a square wave that meets the same condition, but at some other frequency; let curve C be an erratic curve for which there is no name, but which goes through all five points. No doubt the extension corresponding to curve C is ill-entrenched, but this cannot

---

[5] In terms of this context, the analogue of the "grue" example would be a simplified version of the situation described, in which we have examined, say nine samples of (P_0, T_0, k_0) gas and found volume V_0 and "abvolume" A_0, where a thing has abvolume A_0 if either it is one of the first nine cases of (P_0, T_0, k_0) samples examined and has volume V_0 or if it is some other case of (P_0, T_0, k_0) and has volume $2V_0$. The Hullet–Schwartz analysis does not enable us to circumvent the problem of abnormal predicates or extensions, but suggests that it was all along a special case of the curve-plotter's problem.

THE PLAUSIBILITY OF THE ENTRENCHMENT CONCEPT 105

be said for the extensions corresponding to curves A and B. We are in the area where we are dealing with competitions between what Goodman has called "presumptively projectible hypotheses"; hypotheses all of which are supported, unviolated, and unexhausted, and which have survived elimination by both of his two rules designed to eliminate radically unprojectible hypotheses (p. 108). In this area, Goodman tells us, we must not even expect all conflicts to be resolved. Each of "two conflicting hypotheses may be equally valid, with the choice between them depending solely upon decisive further evidence" (p. 108). It is true that he proposes to assess the projectibility of presumptively projectible hypotheses in terms of comparative predicate entrenchment, earned and inherited (pp. 108–117). Presumably on this basis the extensions corresponding to curves A and B will not come off well compared to that corresponding to the extension of "straight line." But to decide whether it is believable that this has anything to do with the superior projectibility of the straight line hypothesis surely requires some discussion.[6] This does not mean that

[6] The curve-plotting problem presents a variety of interesting and relevant features which can perhaps be considered further on another occasion. Here are some: (1) Any intuition that a straight line hypothesis should be favored over a sine wave hypothesis is in the first place an intuition that the set of data points we have is a straight-line-set of data points. That is, we have here an analogue of the judgment that a thing is an emerald and will continue to present an emerald-pattern-of-appearances, or that a thing is green and will continue to present a green-pattern-of-appearances. It is less like the claim that all emeralds will be green because such-and-such emeralds have been green than like the claim that this is an emerald. (2) In practical cases our intuition that a straight line hypothesis should be favored is formed in part by a total empirical context. Given a difference in context, then even despite our expectation that a sine wave property should within five points have revealed itself in some departure from the parallel to the Q-axis, we might expect a sine wave far more than a straight line. Thus if all examined situation S cases have been sine wave cases, then we shall not be content to overthrow the indicated hypothesis with this one case of five "in-line" points. Similarly, if all examined emeralds have been green, one blue-looking emerald is going to be given a longer look before being accepted as such. Goodman's discussion of comparative entrenchment is surely not in defiance of this point, but in abstraction from it. (3) In this connection, where "This is an emerald" or "This is green" appears itself as an induction from non-demonstrative evidence, we might be prepared to entertain the conjecture that not just isolated predicates become entrenched by actual projections, but predicate pairs or complexes. (4) In the discussion above, complications introduced by "inaccuracies" of measurement, or of judgment generally, have been neglected. Where we must allow for a range of error in the determination of our five data points, they may be so placed that a slightly convex curve, never named (and correspondent, therefore, to an unentrenched extension), may be seriously competitive with the straight line hypothesis and much more favored than the sine or square wave hypotheses (despite the far superior entrenchment of corresponding extensions). This is the analog of the situation where we are not sure whether we have an emerald or some other crystalline mineral form. In actual inductive practice our projections reflect the uncertainty consistent with the evidence.

entrenchment considerations never bear on graph competitions, for they do (as the abvolume example shows), especially when inherited entrenchment (pp. 108–117) is taken into account. The present point is that such competitions do not offer strong promise of illuminating the relation between projectibility and entrenchment unless they also involve abnormal predicates. For the study of this relation, therefore, the "simple" "grue" kind of abnormal predicate example may prove to have been a clearer case.

5. We have tried to state reasons why we cannot glibly study the relevance and role of predicate entrenchment in hypothesis projectibility through the use of less abnormal predicates than Goodman has used. Now we must combine with these reasons the consideration that in order to be sure we could, with normal-predicate examples account for the contribution of the abnormal-predicate examples to the analysis of projectibility we apparently would have to make sure what the latter contribution is. This suggests that we must study examples like the "grue" example in any case. For this reason we now turn our attention to the "grue" example itself.

II. The Relevance of the "Grue" Example

1. To state the point of the "grue" example does not require much effort or space, but if the statement is to appear plausible we must first consider at some length the characterizing features of the example relevant to its point. There are three or four features in particular which seem to need mention, because misunderstandings have arisen in connection with them.

a. The first has to do with the reference to time t in the initial characterizing sentence for "grue" (p. 74). This sentence itself does not make clear whether t, as metavariable of the characterization, is instantiable with names or constants designating particular times, or with variables any of which is itself instantiable with constants designating particular times. Nor does it make clear that any such particular time is to be that of some particular inductive occasion. The context, both immediate and remote, definitely establishes that t is to be instantiated with the name of a particular time, a time which is that of a particular inductive occasion.[7]

[7] Support for this is found in Goodman's earlier illustration about the inspection of marbles prior to VE day [5]. It is perfectly possible to mischaracterize "grue" so that the force of this result is blunted. One way is to understand t to be instantiable with any variable ranging over all projection times; i.e., to understand it to apply to anything examined before any projection time just in case it is green but to anything else just in case it is blue. In the characterization of

THE PLAUSIBILITY OF THE ENTRENCHMENT CONCEPT 107

b. The reference to a time t in the characterization of "grue" is relevant to the question of the conditions of applicability of the predicate. Here we have a second feature of the example to discuss. Goodman has framed his example to require that at least by the arrival of the inductive occasion of interest the abnormal predicate may be known to apply whenever the normal predicate is then known to apply. This accounts for Goodman's insistence that we are to compare only hypotheses equally well supported by the evidence on the given occasion (pp. 94, 99–103). The evidence classes for the competing hypotheses must be identical.[8] However, he has also framed the example to allow, though not to require, that the applicability of the abnormal predicate to individual cases may be deter-

---

"grue" the predicate "green" is akin to a variable in the sense that it can connote a multitude of distinct greens. It is not vital to the "grue" example, however, that "green" be such a variable in the example itself. We can treat the "green" of the characterization as a metalinguistic variable and "instantiate" it by supposing it to connote some one very sharply specified hue of green. But if t is to be, under instantiation, a variable in the "grue" example itself, then even where "green" and "blue" are thus sharply specified, "grue" connotes a variety of grues. An entity examined between t_k and t_{k+1} and found green is not grue$_1$, grue$_2$, . . . , nor grue$_k$, but it is grue$_{k+1}$, grue$_{k+2}$, . . . , grue$_n$, where t_n is the time of the present inductive occasion; and so on. By the time t_n, "grue$_j$" would appear in a violated hypothesis for $1 \leq j < n$. Though the hypothesis using "grue$_n$" would be unviolated and as well supported as the competing hypothesis using "green," it could be now maintained that the relation of the "green$_n$" hypothesis to the failed "grue$_j$" hypothesis counted against its predicate's entrenchment and its projectibility. It is this kind of situation, perhaps, that people have in mind who are persuaded that while "green" has applied to past cases and been successfully projected at past times, "grue" has not applied and would not have been successful if tried in projection. (See Sect. III, 2.) Whatever the merits of considering such a predicate it is not Goodman's "grue"; his characterization is precise enough to make that clear. But even if it were not, we would get his hard problem back simply by specifying his characterization as we have already done in the text. In fact the problem can in a sense be made harder than Goodman has made it. To achieve this, we understand t to refer, not to the time t_n of a particular inductive occasion, but to a particular time far removed in the future from t_n. Then, although one can still specify a crucial experiment to decide between this "grue" (no longer Goodman's "grue") and "green," one cannot specify one that will decide before that remote time t.

[8] A thing is in the evidence class for a hypothesis if it is a thing, of the sort described by the antecedent predicate, which is named in a positive instance of the hypothesis, i.e., in an instantiation already determined to be true. A thing is in the projective class for the hypothesis if it is a thing of the same sort not named in a positive or negative case (p. 90). (See also [6].) Incidentally, "grue" cannot be specified as equivalent to "green before t and blue after," for the latter presumably could not be known applicable before t, and if it could, could not apply to just the same cases as "green before t and green after," whose applicability ought to be equally well-known before t (see ns. 9, 18). Hence, "green before t and blue after," though ill-entrenched compared to "green" would not for that reason lose a competition to it, being rather disqualified at the outset.

mined independently of the arrival of the inductive occasion. It is not the case that the point of his example depends on the use of a predicate whose applicability cannot be known until the inductive occasion actually arrives at which it can be seen competing with a normal predicate. As for "grue" itself, if we find an emerald green at a time of examination t' which we know to be earlier than t, then we already know at t' both that "green" applies and also that "grue" applies. And in general there is nothing in what Goodman says to rule out the possibility that the applicability of the abnormal predicate be known as soon as, or even sooner than, the applicability of the normal, if someone can propose a pair of predicates for which this conjecture can be made tolerable. Certainly the conjecture can be tolerated that we may know the applicability of "grue" as soon as we know that of "green," if we already know whether t has passed or not, and otherwise as shortly after as is required to determine whether t has passed or not. Furthermore, if we do wish to characterize a certain abnormal predicate in such a way that its applicability may be known independently of the arrival of a given occasion, we need not mention the time. We may designate the inductive occasion e, or some subsequent event e', in some other way than by its time, or we may tie e (or e') to the number of cases n to be examined before e (e') occurs, or to a designation (naming) of all the individual cases to be examined before e (or e') occurs, or to all the values of a certain property (e.g., pressure) which shall pertain to cases examined on occasions prior to the occurrence of e (e'), or the like. Hence it is a mistake to suppose that on Goodman's characterization "grue" cannot be known to apply (or fail to apply) until time t, or until after time t. It is also a mistake to suppose that if this were so the problem of abnormal predicates would be solved, since we should only have to characterize "grue" as Goodman has in fact done, or in one of the other ways just suggested, in order to have the problem we in fact have.

c. A third feature of the "grue" example has to do with the role of the word "examined" which appears in the initial characterizing sentence (p. 74). The extension of "grue" apparently includes things blue before t, provided there are any such that have not been examined before t, and it does not include all things green before t, provided there are any such not examined before t. The word "examined," therefore, seems to introduce another distinction between grue things and green things besides that introduced by the reference to time t at least if we allow a distinction between being green and being examined and found to be green. Then the extensions of green and grue differ not only in respect to what exists, with what color, after t, but also in respect to what exists, with what color, before t. Accordingly, if we read the characterizing sentence quite literally, we cannot also follow a characterization (A) which says that "grue" applies "to a thing at a given time if and only if either the thing is then green and the time is prior to t, or the thing is then blue and the time not prior to t" [2]. On the other hand, the context of Goodman's introduction of "grue" suggests that he is not building anything important on a distinction between examined green things and green things generally, but rather abstracting from such a possible distinction in the interest of what is vital to a certain kind of case of induction. His purpose is to design a predicate that will catch the evidence class (p. 90) for the projection, at time t, of "All emeralds are green," while at the same time this predicate will have an extension different from that of "green" in a way that would lead to conflicting predictions for unexamined emeralds. For this purpose "grue" might have been characterized to apply (B) to all emeralds examined before t just in case they are green but to other things just in case they are blue, with some augmentation of abnormality, or as in (A) above, to all things existing before t just in case they are green, etc., with some diminution of abnormality. Each of the variant characterizations offers its own small advantages and disadvantages, which need not be rehearsed here. While Goodman's incorporation of "examined" in the characterization of "grue" is not essential to the point of his example, it is essential that the abnormal predicate be characterized so as to be known applicable before t to the cases examined by t.

d. Since the examination of which we speak is intended to delineate the evidence class for a certain hypothesis on a certain inductive occasion, it is clear that "examination" is to be contrasted with "projection," i.e., it is clear that examination definitely establishes something to be the case, while a projection that something is the case is going beyond what has yet been established. This is a fourth feature of the "grue" example that seems to need mention. To decide in a practical situation that a thing belongs in an evidence class, e.g., is green, no doubt takes some measure of what can only fairly be called projection, even when the decision builds on what is quite properly called an examination. But relative to subsequent projection, projection on the basis of this evidence, what is in the evidence class must be contrasted with what is in the projective class. Within the abstraction of the "grue" example, "examination" of a case for a certain property of interest, e.g., greenness, cannot stop with a projection that that case is a case of greenness. Hence the example must not be understood in such a way that the examination of a thing can be found at odds with a projection for the thing. What has once and for all been established by examination is established; it cannot be in conflict with what is projected, which is what has not yet been established; neither can it be in conflict with the results of a subsequent examination, which surely cannot set aside what has already been established (or else "established" was incorrectly used in the first place).

Any objection to meeting this requirement can at best be a claim that it is unrealistic to suppose that it could be met in empirical practice. But while this is true, it is largely because it is unrealistic to suppose that in empirical practice we could establish the color of a thing (any more than the abcolor) right through t by an examination (completed) at t'. This point is confused by linguistic practice, in accord with which we do not honor with our predicates a sharp distinction between what is established and what is only projected. Our normal and "realistic" use of "is established green" or "is found green" applies such expressions to things like emeralds and not things like emerald-temporal-segments, and is projective.[9] If we try to adhere to this use, we must be careful to avoid begging the question. That is, if we construe the "grue" example as though emeralds could at t' prior to t be examined and established as to color through examination time t'' subsequent to t, then we must allow that the same can be done for "abcolor." In particular, we are not free to construe the example to allow a second examination of an emerald, or at least a second examination issuing in an assignment of "grue" and "not grue" to "the same emerald." Furthermore, suppose we examine an emerald and establish it to be blue for a time interval that ends before t. We nevertheless presumably establish for a time interval that extends indefinitely beyond t that this emerald

[9] If we speak of "establishing" an emerald to be green, without any mention of the time at which and over which it was established to be green, we almost invariably are making a tacit projection over the emerald's past and future as an emerald. That is, we certainly have not in fact established, for example, that the emerald has been and will be green at least for so long as it has been and will be an emerald. What we have established at most is that a relatively small temporal segment of the emerald in question, perhaps inclusive of, or overlapping, the temporal segment which underwent examination, is green. In fact, even this much is doubtful, when we consider that we must allow for the distinction between "is green" and "looks green," and for the fact that the very application of the former probably generally involves projection. (See n. 6, comment 1.) The implications of the fact that our "realistic" use of "is established green" is partially projective are further developed in Sect. IV, 1. In particular, compare n. 18.

was examined before t. This fact is relevant, given Goodman's characterization of "grue" as I have represented it above, to the applicability of abcolor predicates. If we allow a second examination after t, we shall face perplexities unless we are careful to construe the examination of a thing for abcolor as the characterization requires. That is, the abcolor examination must have the same effect as if, in addition to examining a thing for color we examine it also for examination-prior-to-t (see [15], p. 530, and [10], p. 270). However, it is likely that Goodman's original characterization has in view the simplifying assumption that examining an emerald for color (or abcolor) establishes its color (or abcolor) indefinitely. Goodman's evidence class seems full of emeralds established green, not emeralds established green over some limited temporal interval.[10]

2. We have discussed four features of the "grue" example that are relevant to its point but have been foci of misunderstanding. We now want to emphasize that what is essential to this example, or any other designed to make the same point, is that the predicates represented as competing at a given projection time t must differ in applicability, but only over cases not yet examined at t (in the above sense). Thus the extensions of "grue" and "green" must differ, but

---

[10] Of course, we can avoid examining things for examination-prior-to-t by recharacterizing "grue" appropriately ([15], p. 531). Independently of that point, we can decide that it is in fact more realistic to apply "green at time T" to emeralds than to apply "green," or to apply "green" to emerald-temporal-segments than to emeralds. Such moves, however, bring along their own unfamiliarities. One can get emerald-temporal-segments for predicate extensions through a rather ordinary mode of speech, by attributing "green" or "grue" to things only at given times, e.g., at times when they are examined and found green (or grue) or at times when they are green (or grue). This approach has been followed by several of Goodman's commentators, including Skyrms ([13], p. 57) and Barker and Achinstein ([2]; see Sect. II, 1c). This device does not necessarily eliminate infringements of ordinary locution, however. If "emerald" in (i) "All emeralds are green" denotes emeralds, then establishing that emerald a is green at a given time T is evidently not to establish that emerald a is green. Hence we do not have a positive instance of the hypothesis in question, although we do of the hypothesis (ii) "All emeralds are green at time T," and even of (iii) "All emerald-segments are green." Hypothesis (ii) would be a far-fetched reading of (i), and to get to (i) from (ii) would incorporate further projecting, the analysis of which would require dealing with a variety of predicates differing in specification of T. (We cannot, incidentally, understand "Emerald x is green at T" to mean "Emerald x is a green thing and an exists-at-T thing," at least without a convention that "green" applies to x so long as "exists-at-T" applies, but not necessarily longer. For otherwise "green" in "Emerald x is a green thing" would in general be understood to apply longer, perhaps as long as "emerald" applies.) Hypothesis (i) can be regarded as entailed by (iii), and so confirmed by the evidence about a at T, but then the confirmation is not direct in Hempel's sense [9]. If we want direct or "instance" confirmation, then "emerald" in (i) must be read as denoting emerald-segments.

anything, emerald or something else, already examined at t and established to belong to either extension must also belong to the other. Nothing literally established by projection time t, as distinct from what may have been conjectured by prior projection to hold beyond t, may distinguish a case as satisfying one of the competing predicates but not the other. In terms of their descriptive adequacy to the empirical instances available at t we must have no basis to choose between the two predicates, while yet their conditions of applicability guarantee that over some cases as yet unexamined not both can apply.

Having said this, we can now say what the point of the "grue" example is. This is to illustrate that our intuitive projection preferences cannot be explained, nor a justifying principle of comparative predicate projectibility constituted, solely in terms of the comparative descriptive adequacy of predicates to what has been literally established about the evidence cases.

3. If this point is tentatively accepted, then the question arises: What is required of a predicate besides adequacy to the evidence? To this Goodman has made the reply that adequate entrenchment is needed as well. One may refuse to accept this point, of course. If he does so, he presumably must return to the attempt to explain the superior projectibility of "green" over "grue" hypothesis in terms of a difference in their descriptive adequacy to the data available at t. One immediately thinks that the example, along with the confirmation theory to which it is oriented, may exhibit too narrow a conception of available data. Should we not seek a more realistic conception, perhaps even take account of how certain predicates have fared in projection, or would have fared had they been tried, perhaps even consider a theory of predicate generation? There are three things we should remember if we adopt this approach.

a. The appeal to entrenchment is explicitly an appeal to a wider conception of data, in fact to data about the relative success of predicates in projection (pp. 85-87, 92-94). We cannot suppose that every useful predicate has become well-entrenched, nor that those which have could not be dispensed with in favor of others never tried. Neither can we suppose that a given well-entrenched predicate has appeared only in unviolated hypotheses, but we can be sure it has not appeared only in quickly and repeatedly violated hypotheses. The success and failure of prior projections "underdetermine" the selection of our stock of well-entrenched predicates, but in part they do determine it.

b. A second point to remember is that Goodman's analysis explicitly endorses our appreciation that predicate selection is in part

THE PLAUSIBILITY OF THE ENTRENCHMENT CONCEPT 113

constrained by empirical considerations other than those having to do with how many times the predicate has been projected (entrenchment). This is, after all, why only supported, unviolated, and unexhausted hypotheses are considered in the two elimination rules (pp. 94, 100, 108). However well-entrenched a predicate may be, a violated hypothesis using it is not found projectible. However ill-entrenched a predicate may be, a supported, unviolated, unexhausted predicate using it is not found unprojectible if it has no competitor meeting the same criteria and using a better-entrenched predicate (p. 97). Hence to be projectible at all a hypothesis must in the usual way answer to the way the world extralinguistically is, and in certain cases a hypothesis which does so is projectible even if it uses very poorly entrenched predicates.

c. Finally we must remember that by his own claim Goodman's primary interest is in isolating a reliable indicator of projectibility (p. 98). His fundamental hypothesis is that superior entrenchment is a necessary and sufficient condition for a valid projective preference (pp. 98, 108 ff.), and hence may be cited as a justification for a particular projective choice (pp. 64, 65).[11] Hence an entrenchment solution is not necessarily incompatible with a solution in terms of still other data; perhaps, for example, data founding a particular theory of predicate generation and retention in a language.

The latter point is equally true if one accepts the point of the "grue" example, but seeks a solution in either formal or psychological terms, e.g., in terms of formal or psychological "simplicity," ease of applicability, or the like. However, some of these proposed solutions appear no more promising, under investigation, than do proposals which refuse to accept the point of the "grue" example. In fact, when one considers these alternative explanations of projectibility judgments,

[11] "Hypothesis" may not be the very best word, but it will keep us aware that Goodman does not attempt to establish a projectibility-indicator role for entrenchment by arguing that an explicit justifying appeal to superior entrenchment is the characteristic feature we find in what we regard as samples of reliable projection. His suggestion that the "judgment of projectibility has derived from the habitual projection, rather than the habitual projection from the judgment of projectibility" (p. 98) must not be stretched that far. Incidentally, the latter suggestion seems to provide ground for regarding superior entrenchment as necessary for superior projectibility. Goodman speaks as though he would be satisfied to have superior entrenchment provide merely a sufficient condition (p. 98). This may be on the analogy of deductive rules, each of which is sufficient but not necessary for deductive validity. But if we may have superior projectibility without superior entrenchment then we must have a supplemental rule to help us recognize a case to which entrenchment comparisons apply. It is true that Goodman's elimination rules are explicitly circumscribed in their application, but entrenchment comparisons are meant to apply more broadly (pp. 106, 118, 119).

of whatever sort, many of them turn out to be inadequate, or else too vaguely programmatic for further evaluation. Both the "grue" example and the entrenchment solution, on the other hand, gain strength in the comparison. For this reason we first consider a sample of these alternative proposals, including a number already discussed by Goodman either in [4] or in a reply to one or another objecting article. After that we shall try to allay discontent with the entrenchment solution itself.

III. ALTERNATIVE SOLUTIONS TO THE PROJECTION PROBLEM

Our interest in alternatives to entrenchment is not in classifying them. We shall not attempt to distinguish evidential, formal, or psychological solutions, nor mixtures of these. Similarly no significance attaches to the order of their consideration. For convenience, and because it will help us to distinguish proposals building on mere idiosyncrasies of the "grue" example, we draw up most proposals in terms of that example.

"All emeralds are green," we may say, is to be preferred to "All emeralds are grue," because:

1. "Green is a more readily applicable predicate than "grue." The idea here would be that to apply "grue" one must meet every condition needed to apply "green" together with some other.

This proposal seems to require an analysis and formalizing of condition-stating. Even if we grant "grue" were in the stated terms less readily applicable than "green" would we be safe in eliminating it for that reason? Suppose G, "looks green under green light," applies to all examined emeralds. There are many things to which this applies that "green" does not; to apply the latter seems therefore to require meeting more conditions. Does it also require meeting more conditions than to apply R, "looks grue under grue light," which would perhaps be said to require meeting one more condition (a check of time or date) than G? If so, then should we project R in preference to "green," which surely conflicts with it over unexamined emeralds?[12]

2. Predicates are proposed in order to try them against future cases, and "grue," had it been projected at earlier times than the present occasion of inductive choice, would immediately have failed subsequent tests prior to the present occasion, whereas "green" would have been successful.

[12] Even G competes with "green" over unexamined emeralds in the sense that subsequently examined emeralds might satisfy the former but not the latter.

Unfortunately this line of reasoning has in fact missed an important implication of Goodman's "grue" example. This is that if "grue" ever had come into actual trial in projection instead of "green" it would have been projected as successfully as "green" at all projection times prior to t. The t in the characterization of "grue" refers us to some particular time, let us call it t_n, which is that of a particular inductive occasion at which both "grue" and "green" are seen to apply to all the same things (see Sect. II, 1a above). On any prior inductive occasion, of time $t_k < t_n$ the projection of "grue" would have been found successful in, for example, the next examined case, provided the projection of "green" would have been found successful in that case, since this case would be a case of something examined before t_n and green. Neither the nonentrenchment of a predicate, nor its nonprojectibility (if we distinguish them), tells us that it would not have been successful had it been tried previously.

3. The generation of a predicate is psychologically in the nature of a response, at least in part, to "what is presented." What has not yet been experienced elicits no response, in word or otherwise (although later it may prompt responses now unimaginable). The "distribution" and "intensity" of language-eliciting factors or aspects in what has been experienced is reflected in the predicates we propose. Predicates are not mere words aimlessly created in a vacuum, unshaped by any external empirical influence, the product of a random predicate generator. "Green" reflects the way the world is in its eliciting of predicates. "Grue" should be eliminated because, although it would be successful if tried, its coming to trial requires a causal sequence that has not been realized in our world. It requires that at some time t' prior to t (a more or less contemporary projection time), time t has somehow made its "importance" felt in a way that calls for a covering predicate; and this is almost certainly not the case in fact (and is if anything the more unlikely, the greater the interval between t' and t), though as a matter of sheer logical possibility it is not excluded. Similarly any predicate should be eliminated whose conditions of applicability involve an appeal to any factor which cannot have causally provoked, at some time or other, a particularly suited linguistic response.

This criterion offers some promise of application in terms of a "fitting" of words to extralinguistic features. However, it does so at the expense of an unrealistic theory of linguistic behavior. Let it be granted that in order actually to have used and projected "grue" with significance at any time, early or late, we must have known at that time approximately under what circumstances to apply it or not to apply it. Still a predicate is proposed not just to answer past experience, but to answer future experience as well. One of the features of human "creative" thinking is that it involves the synthesizing of what has been analyzed as elements in the naive unity of experience into groupings, linkages, and patterns never actually experienced as such. This feature appears in projective behavior when for example, crucial experiments are proposed to decide between hypotheses. Who is to say that "in the early days" of "the mind's" trial of predicates (p. 87) some future time t may not have figured in someone's synthesis, with which he intended to unify future and past experience? It is not inconsistent with even a causal theory of language generation to say this, but only with a narrow and unrealistic causal theory. On the basis of a better theory, it is not at all clear that "grue" might not have been tried.[13]

To be sure, the point of the criterion we are considering may be made immune to the objection of the preceding paragraph by proposing a better theory. We need only ask why we may not seek the recurrent extralinguistically oriented characteristics of any or all predicates we would be caused to try, however we might be caused to try them, and also the characteristics of any that may be mentioned which we would not be caused to try, and certify projectibility in terms of the former. But about this proposal three things may be said. First, it is not yet a criterion, but only a proposal for one, since it depends on the promulgation of a successful causal theory of predicate generation. Would we, or would we not, be caused to try "grue"? Secondly, although it is a proposal for an extralinguistically oriented criterion, we cannot suppose in advance that it is a proposal for a criterion in which predicates are preferred solely because they provide a superior "fit" with extralinguistic phenomena. The predicates preferred are to be those which could be consequent upon empirical states of affairs, but we cannot suppose that in view of this they will be predicates which better "picture" or "answer to" those states of affairs. Thirdly, we cannot suppose in advance that we shall even be able to achieve a criterion drawn wholly in terms of extralinguistically oriented characteristics of predicates. It is logically possible that initial projections would appear as consequent upon

---

[13] It may be worth remarking that Goodman's speaking of "the mind as in motion from the start, striking out with spontaneous predictions in dozens of directions" (p. 87) of course does not imply an a-causal theory of predicate generation. The common conviction that language has conventional (or happenstance) elements may be understood as reflecting something about the facts of actual language generation for which a sophisticated causal theory must account.

THE PLAUSIBILITY OF THE ENTRENCHMENT CONCEPT 117

extralinguistic phenomena but later projections as caused partly by linguistic phenomena in such a way that only by reference to such linguistic phenomena could we distinguish a recurrent set of characteristics to be correlated with projectibility. Entrenchment, of course, would be a candidate for just such a linguistic phenomenon.[14]

4. The definition of "grue" involves a reference to time, place, or finite number of individuals.

As Goodman has pointed out (p. 80) this is true of "grue" defined in terms of "green" and "blue"; it is also true of "green" defined in terms of "grue" and "bleen."

5. The semantic meaning or significance of "grue" involves a reference to time, place, or finite number of individuals. Insofar as this is a suggestion different from the previous one it must have to do with a distinction that can be made between "grue" or "green" in terms of what we would actually have to do to come to employ the terms, or to understand them, or to apply them satisfactorily. The first of these has been considered under suggestion 3 above. The kind of reply that could be made to the others is indicated by what has been said under 1 above.[15]

---

[14] In the absence of an adequate scientific theory of predicate generation, our present persuasion that a predicate is not one that would have been tried (and successfully so) could lie in the fact that it is not entrenched. This at least is what Goodman is saying.

[15] Goodman's discussion of "qualitativeness" (pp. 79, 80) is not drawn in terms of a distinction between a syntactic (e.g., proposal 4) and a semantic criterion (e.g., proposal 5) of qualitativeness. I am inclined to suppose he is addressing himself primarily to something like 5 as the stronger of the two, and I take him to be making at least two points. The first is that it is doubtful if an independent criterion of qualitativeness can be achieved which does not beg the projectibility question. (As Hullett and Schwartz observe, [10], pp. 264–266, if we suggest a certain criterion for distinguishing "purely qualitative" predicates, we still have to make plausible the hypothesis that such predicates must always be the ones favored in a projection decision, which seems, in fact, unlikely.) Goodman's second point is that while we can indeed specify a concept of qualitativeness with the help of the particular sort of time-involving characterization he has given for "grue" and "bleen," this concept is not absolute but relative. That is, if "green" and "blue" are temporally qualitative predicates, then "grue" and "bleen" must be nonqualitative. But since "green" and "blue" can be characterized in the same sort of way, it is also true that if "grue" and "bleen" are temporally qualitative, then "green" and "blue" must be nonqualitative. But these facts can ground only a judgment of relative nonqualitativeness, unless supplemented with an independent criterion of absolute qualitativeness. Fain's recent discussion of qualitativeness [3] seems to presume what Goodman doubts, that such an independent, absolute criterion can be achieved which will not beg the projectibility question. Furthermore he seems to presume that according to some such criterion "green" is obviously qualitative (and of course that "grue" is not). Fain says it is as absurd to call "green" temporally nonqualitative because it can be defined with the help of a "temporal" expression as to call "red" self-contradictory because it can be defined with the help of a self-contradictory

6. "Green" is simpler than "grue."

In what sense simpler? If graphical simplicity is intended, presumably what is meant is that a plot against time of emeralds examined for color looks simpler if all the emeralds fall on the line representing the color green (straight line), than if those examined after t fall on the line representing blue (step-function). This is true, but similarly a plot of emeralds examined for abcolor looks simpler if all the emeralds fall on the line representing the abcolor grue (straight line), than if those examined before t fall on the line representing grue while those examined after t fall on the line representing bleen (step-function). The criterion does not help us to decide between examining for color and for abcolor.

But perhaps a criterion of psychological simplicity, e.g., of applicability, or of response to what has reached us in experience, is intended. Such a suggestion would not imply a necessary conflict with a projectibility criterion, as already noted. But the kind of problems that might arise in connection with it have been indicated under proposals 1 and 3 above.

7. The hypothesis using "green" is related to other hypotheses which contribute to its endorsement; not so that using "grue."

This is true as a matter of fact, but again it does not help us to distinguish "grue" from "green" except by the fact that "green" is in use in various hypotheses, some of which mutually support one another, while "grue" is not. Depending on one's understanding of the essential features of plausible endorsement mechanisms (pp. 107–118), one can even make a case for the continuance of all endorsement relations under the uniform substitution of "grue" for "green." However that may be, it seems that for a given hypothesis h endorsing

---

expression. This is true but not at issue. Whatever the merits of arguing analogically from the actuality of the "semantic self-consistency" (my term) of "red" to the possibility of the "semantic qualitativeness" of "green," it is the actuality of the latter that Goodman doubts. Fain's contention that qualitativeness is no more a relative matter than contradictoriness depends on our being able to tell whether a given predicate is semantically qualitative as we can tell whether it is semantically self-consistent (supposing the latter granted). In this connection it is worth noting that Goodman ([7], p. 330) has argued against Thomson ([14]; see also her reply [15]) that "grue" can be applied correctly without meeting the alleged extra condition of applicability, i.e., knowledge of the current time, which would presumably be the sort of thing Fain has in mind as definitely establishing the semantic nonqualitativeness of "grue" in contrast to "green." (Ackermann's interesting discussion in [1], pp. 30–32, is relevant here, and the more evidently so if we modify his example to allow his specification of "grue" to approximate more closely Goodman's original characterization.) Once again, however, the more important question is whether in any case the specification of such an extra condition of applicability can be generalized to capture all and only nonprojectible predicates without question-begging.

the hypothesis using "green" one can devise hypothesis h', competing with h, which endorses the use of "grue," the competition being of the sort already at issue (pp. 76, 77).

8. If the characterization of "grue" in terms of "green" is relativized to any inductive occasion prior to the current one whose time is t_n, the time t becomes that of that prior occasion, t_k. Then the projection to the next case is in disagreement with the results of the already past examination of that case. We take as our criterion that no predicate is to be allowed which cannot pass this relativization test.

This criterion appears promising at first, because it seems to distinguish between "green" and "grue" in favor of the former. Even if "green" is characterized in terms of "grue" and "bleen" and t, when this characterization is relativized to a prior inductive occasion of time t_k the projection to the next case agrees with the now past examination of that case; i.e., that case is one examined after t_k and green. But while this is true, it is true only if we relativize the characterization of "grue" and "bleen" under the constraint that "green" and "blue" retain the extensions they have in present usage. On what non-question-begging ground can we justify using a test which thus favors the normal predicates? Suppose instead I consider "green" to be the defined term and insist on a relativization of its characterization (i.e., taking 't_n' into 't_k') under the constraint that "grue" and "bleen" retain the extensions given them by Goodman's actual characterizations (in terms of 't_n') of them. Then the projection of "grue" at t_k would have been shown successful at the next examination and that of "green" (as relativized) a failure.

Furthermore, the proposed criterion has two other deficiencies.[16] The first is that it cannot be satisfactorily generalized in any obvious way. We realize at once that the unprojectibility of unprojectible predicates cannot always be tied to the time of the examination of evidence cases. Suppose we introduce "grupe" to apply to those things examined at place p and found green or to other things found blue. Then if all our evidence cases up to the time of the present inductive occasion have been examined at p, a relativization in terms of time of inductive occasion will clearly leave "green" and "grupe" on equal footing. Can we then eliminate "grupe" by relativizing in terms of place of examination? All evidence cases have in fact been examined at place p. How can we say what would have been found

[16] In a preliminary version of this paper these were the only two deficiencies noted. The "asymmetry" between "green" and "grue" established by the proposed criterion I took at face value until a questioning comment by Professor Goodman led me to think further about it.

true of one of them had it been examined at some other place without begging the question at issue on the present inductive occasion? It is clear that while projection may depend on a time difference between examined and unexamined cases the "grue" example does not get its point by any unfair exploitation of that fact.

In the second place, the present criterion, even in the restricted version we are considering, seems to throw out predicates we might very well want. There are surely situations in which we want to predict, for example, that some alteration in phenomenal aspect of certain entities will take place at such and such a time. In a competition to describe these entities, and project over them, a predicate will surely be preferred that cannot pass the relativization test.

9. Consider two competing hypotheses, H and H', where H uses a predicate whose conditions of application would not prevent a crucial experiment's having been carried out prior to the time t of the present inductive occasion for every hypothesis using the predicate, while H' uses a predicate whose conditions of application would prevent such an experiment in the case of every hypothesis using that predicate and formed to compete with some hypothesis using the alternative predicate; H is to be projected rather than H'.

The idea here is plain enough. For no hypothesis pair using "green" and "grue" respectively does it appear that one could manage a test case to decide between the members of the pair prior to time t. What is not so plain is how we can specify which predicate's conditions of applicability are at fault without prejudice in favor of the one that happens to be in use, namely "green." For we must preserve the relationship between "green" and "grue" determined by Goodman's characterization of the latter, or else we have altered the example we are trying to explain. Hence "green" and "grue" must apply equally well over all examined cases of all hypotheses using either before t, and disagree over all examined cases after t. How is it any more the fault of the conditions of applicability of "grue" than of those of "green" that they should not be able to be found in disagreement before t, except that "green" was here first? It is quite true that on the inductive occasion of time t we expect future examined cases to be "continuous in greenness" with past examined cases, and to be "discontinuous in grueness." But this agreement on the one hand, and this disagreement on the other, are not at time t matters of "descriptive knowledge," but of expectation, and neither can be used at time t to count for, or against either hypothesis. Had we been all along thinking of green things not as such but as grue things and been expecting them to be continuous in grueness, then we should be inclined to blame the conditions of applicability of "green" for our not being able to decide between "green" and "grue" before t.

Hence it appears that this criterion cannot discriminate projectible from unprojectible hypotheses. But even if it could, and if it succeeded then in rejecting only unprojectible hypotheses, it could not reject all unprojectible hypotheses. For it may be that no cases of green things examined by time t happen to be of class B, so that we can think to project "All emeralds are green" and "All emeralds are grube," where "grube" applies to anything not in class B just in case it is green, and to anything in class B just in case it is blue, and where nothing in the conditions of applicability of "green" or of "class B" (and therefore of "grube") precludes that something of class B has been examined before t. (E.g., class B might be the class of things located at any of places $P_1, P_2, \ldots, P_i$, or the class of things broken in two by any hydraulic press of a certain sort, and so on.) But "All emeralds are grube" is no more projectible than "All emeralds are grue," by comparison with "All emeralds are green."[17]

IV. COATING THE ENTRENCHMENT PILL

Certainly other proposals might be considered, and perhaps eventually one of them may be made to work, either ruling out the entrenchment proposal or "explaining" it in other terms. But the negative results just noted make it worthwhile to give the entrenchment proposal a sympathetic hearing. This means, in part, giving credence to the possibility that the intended point of the "grue" example is well taken, i.e., that we cannot explain intuitive projection preferences, nor ground a justifying principle of comparative projectibility, in the comparative descriptive adequacy of predicates to what has been literally established in "the evidence." But it also means what may for some be harder to allow, entertaining the possibility that what is needed in a predicate besides descriptive adequacy is a sufficient degree of comparative entrenchment. In the remainder of this paper we try to make it easier to entertain both these possibilities by relating them to other, presumably familiar, features of our epistemological response to experience. We do so first by arguing that we have a familiar way of ascribing properties to things, reflected in the usual significance of our predicates, which

[17] A variation of proposal 9 would rule out any predicate "disjunctive" in conditions of applicability, one of whose disjuncts is empty of examined cases at t. Unfortunately, "examined after t and bleen" or "examined after t and green" alike constitute such disjuncts.

goes beyond what has in fact been established in our experience and is therefore in part anticipatory; if we take note of this fact, both possibilities become more believable in a rather obvious way. Secondly we argue that the entrenchment proposal is quite credible as an answer to the need suggested by the point of the "grue" example because it can be seen as a special case of a methodological principle employed in answer to similar needs fairly general both in science and in our everyday response to our environment. To these two arguments we now turn.

1. Goodman observes that Kant has taught us to be wary of supposing that "the way things are" is independent of our knowing that they are that way. Then he suggests that what we mean by "the way things are" at any given time is in part determined by the words in use in our language, and that it is just to this part that we must have recourse to explain our inductive preferences. This seems hard to take because it seems to require that things turn out to be green rather than grue because we happen to have started applying "green" to things rather than "grue." But no matter how many emeralds may have been found to be grue up to this time, we simply don't believe that those examined after t will be found grue. And this, we feel, has nothing to do with the mere fact that "grue" has never been actually projected before, because that has nothing to do with the question whether the emerald next examined after t is blue or green.

However, since Goodman explicitly denounces the attempt to decide the validity of projections by reference to unexperienced cases (p. 99), we should seek an understanding of his suggestion without this unacceptable implication. To say that the way things are is partly determined by the words we use would be truistic if it meant merely that what we expect to experience in the future is evidenced in the predicates with which we project over future cases; and it would be misleading (if not worse) if it were only an alternative locution for a recommendation that we continue to use the predicates in successful use. A better understanding of the suggestion is forthcoming, I believe, if we consider what lies behind our utterance when we insist that things just "are" green and just "are not" grue. There is a common and legitimate use of "are" which will tolerate the construction of such an utterance. Indeed when we say, "the way things are," we are almost always making such a use. It is a use which incorporates a projective element. In my judgment, it is of this "are" which anticipates what shall be, and does not just reflect on what has been literally established, that a predicate-in-use can be understood as partly determinative.

To revert to our example, if we insist that despite the applicability of "grue" to all and only those tried cases to which "green" applies nevertheless things just aren't grue but they are green, we must be using "are" in a way that anticipates untried cases. If we were talking only about tried cases we would be flatly contradicting ourselves, since the characterization of "grue" guarantees its application to the same tried cases as "green."[18] In this sense of "are" where we are already talking about cases in general, and about what exhaustive examination alone can decide, and not just about what can be affirmed on the basis of cases actually tried, the way things "are" depends in part, as Goodman says, on "how the world ... has been described and anticipated in words" (p. 119). Not with the careful, worried "are" of epistemological discourse, in which we distinguish what has actually been established from what may perhaps be found to be, do we say that the way things are depends on the predicates we use, but with the generous, trusting everyday "are" that reports our persuaded anticipation about things as a facet of a single insight into their observed nature. What we are willing to concede to hold for all examined emeralds, namely that they are both grue and green, does not depend on the words we use, but on what has been presented. What we are persuaded holds both for examined and unexamined emeralds, namely that they are green, depends not just on what is presented, but on the words that have been used, and used successfully, in "organizing" the presented (p. 96). The usage of words is partly determinative, not of truth, but of projective validity.

2. But we have yet to say how it can be satisfactory to regard superior entrenchment of a predicate as the normative mark of

---

[18] Exactly the same can be true even where the past tense is used, as in "Emeralds have been green rather than grue," and even where there seems to be an explicit restriction to tried cases, as in, "All tried cases of emeralds have been green rather than grue." In the respect in which an emerald would normally be considered a tried case, the emerald has not been tried through the time t of the present inductive occasion. (See Sect. II, 1d, and n. 9.) The trial, therefore, cannot have distinguished any emerald as green *rather than* grue. To speak in this way is to take the actual trial to have established something about an emerald over what is, by hypothesis of the "grue" example, the temporal span of the projective. It is to anticipate that future scrutiny will show already scrutinized emeralds continous in greenness through t, and is therefore already to make a projection, despite the past tense. Of course we can, for simplicity, assume that trial of an emerald establishes its color through t, whence it must, by the hypothesis of this assumption, be wholly in the evidence class of emeralds at t and wholly excluded from the projective class (n. 8). But since fixing color behavior through hypothesis entails a concurrent fixing of abcolor behavior, it remains true that pre-t tried emeralds cannot be green and not also be grue. The same holds if we take "tried case of emerald" to mean, also unnaturally, what is meant by "exhaustively scrutinized emerald-pre-t-temporal-segment."

superior projectibility of a hypothesis using that predicate. The only answer that can be given, it seems to me, is that nothing in the way of a superior fit to the data would be provided by an alternative choice of (a less-well-entrenched) predicate at the present projection time, and nothing in the way of superior projective promise. "Green" has done for us, in all examined cases, everything that "grue" could have done for us, both descriptively and predictively. To see things as grue rather than green is admittedly logically possible, but gains us nothing that we do not already have in seeing them as green. Where a newly suggested predicate differs relevantly from the old only in untried cases of every hypothesis in which the old has always appeared, what would prompt a switch to the new?[19] To put it another way, we don't consider projecting a new predicate without a good reason, and the wholly satisfactory service of an old predicate is not a good reason. Our reason for preferring "green" to "grue" at the present projection time is that we have no good reason to prefer "grue" to "green."

From a logical standpoint there is no doubt that this amounts to a kind of conventionalism. However it is the same sort of more or less virtuous conventionalism that operates in much of scientific methodology, particularly in our reliance on hypotheses that have proved useful. We do not set one hypothesis aside in favor of a competitor that does nothing for any existing evidence whatsoever that the first does not do (i.e., without a reason in the empirical situation).[20] May we not become accustomed to the thought that we do not set one predicate aside in favor of a competitor that does nothing for any existing evidence whatsoever that it does not do? To choose a better entrenched predicate in a projection seems to be merely the following of an extension of the same methodological principle of insufficient reason used in the selection of a better established hypothesis over alternative hypotheses. We have seen this principle employed in the latter way long before Goodman showed us new ways of generating such alternative hypotheses through predicate tailoring.

[19] "New" here refers us to predicates with a poor record of actual projective use. There is no question there may be reason to try in a certain area predicates new to that area, but otherwise old. For example, cases of A, always found to be B, may come to be seen as special cases of A', which have always been found to be C. This could motivate the projection that every A is a C also, though only in as-yet-unexamined cases of A could the difference between B and C be determined, and though 'C' has never been applied to an A before.

[20] Even in the case of broad, integrative theories, attractive on non-empirical grounds (e.g., scope, simplicity), a competing older theory is not regarded as empirically inferior (e.g., "merely approximate") unless an empirical basis for this devaluation is found.

What we are saying, after all, amounts to this, that a predicate is a part of what is being tried against the world when we try the hypothesis in which it appears against the world. It is being tried for its success in projecting as the hypothesis itself is being tried for its success as a projection. Indeed, in a certain sense a predicate is a kind of hypothesis. For as I tried to argue above, even to say that a particular thing of a certain kind is green is quite commonly to mean that it is not, among other things, grue. But it makes no sense to regard such an "excluding" assertion merely as a statement of fact. It is a tacit hypothesis about what shall be found to hold about other things of the same kind. It is a statement whose function has gone halfway towards that of the generic particular form, more readily recognizable as hypothesis, "The emerald is green" (compare "The tiger is a mammal").

Of course, there is a difference in the estimation of the success of a given predicate and of a given hypothesis employing it, because the two are not held to the same criterion of utility. This is only to repeat what has been said earlier. The failure of the given hypothesis does not by itself annul the predicate's success-in-projecting character. This is true if the predicate has been often successfully projected, perhaps even in this hypothesis. It is true also if the predicate has never been tried before, or tried but never successfully, for that matter.[21] Similarly a hypothesis may succeed in one or more trials, while the predicate it uses is one that has been remarkably unsuccessful in other hypotheses and may shortly turn out the same in the present one also.

Nevertheless, in a general way, predicates are tried for success-in-projecting as the hypotheses in which they appear are tried for success as projections. As an unviolated, confirmed hypothesis is only one of a great number of alternatives which are equally well supported by the same positive instances, so its entrenched predicate P is only one of a great number of alternatives which equally well apply to all the same entities described by all positive instances of all the hypotheses in which P has appeared. To rely on predicate entrenchment in the normative assessment of projectibility seems neither more nor less

[21] Incidentally, were not entrenchment a relatively coarse measure of projectibility anyhow, one would perhaps want to define degree of entrenchment in terms of number of successful projections, rather than in terms of mere projections, and perhaps even introduce positive and negative entrenchment. Ignoring the distinction is warranted, of course, by the consideration that for a well-entrenched predicate there has presumably been until now an average temporal convergence of number of successful projections toward total number of projections.

harmful than to rely on previously satisfactory hypotheses generally.

The present discussion brings us full circle to the problem of Sect. I. May we reduce the problem of alternative predicates entirely to that of alternative hypotheses, or vice versa? As promised in Sect. I, this is a question we consider no further here, along with such as the following: How does the firm violation of a hypothesis, which certainly necessitates a new hypothesis, stand in respect to the necessitation of new predicates? Can we count on relative predicate entrenchment to mark relative psychological simplicity in the reactions of "the human race" to the presented? Do we invoke new predicates (new extensions) not just because of hypothesis violation but also to serve alleged hypothesis broadening to cover more evidence from diverse areas? The discussion of these questions may be allowed to wait another opportunity.[22]

University of Pennsylvania

REFERENCES

[1] Robert Ackermann, *Nondeductive Inference* (New York, Dover, 1966).
[2] S. F. Barker and Peter Achinstein, "On the New Riddle of Induction," *The Philosophical Review*, vol. 69 (1960), pp. 511–522.
[3] Haskell Fain, "The Very Thought of Grue," *The Philosophical Review*, vol. 76 (1967), pp. 61–73.
[4] Nelson Goodman, *Fact, Fiction, and Forecast* (Cambridge, Massachusetts, Harvard University Press, 1955; 2d ed.; New York, Bobbs-Merrill, 1965). (Page numbers appearing in parentheses in the text are references to the second edition of this work.)
[5] ——— "A Query on Confirmation," *The Journal of Philosophy*, vol. 43 (1946), pp. 383–385.
[6] ——— "Faulty Formalization," *The Journal of Philosophy*, vol. 60 (1963), pp. 578–579.
[7] ——— "Comments," *The Journal of Philosophy*, vol. 63 (1966), pp. 328–331.
[8] Marsha Hanen, "Goodman, Wallace, and the Equivalence Condition," *The Journal of Philosophy*, vol. 64 (1967), pp. 271–280.
[9] Carl Hempel, "Studies in the Logic of Confirmation," *Mind*, vol. 54 (1945), pp. 1–26, 97–121; reprinted in Hempel, *Aspects of Scientific Explanation* (New York, Free Press, 1965), pp. 3–46, with a postscript written in 1964, pp. 47–51.

[22] Work on which this paper is based was supported in part by stipend of a Mellon Postdoctoral Fellowship at the University of Pittsburgh, 1966–67. I should like to thank Robert Schwartz, Robert Ackermann, and Professor Nelson Goodman for reading this paper evaluatively. As a consequence of their comments and questions a number of modifications have been incorporated into the text in the interest of a clearer and more correct statement of "grue" problem complexities.

[10] James Hullett and Robert Schwartz, "Grue: Some Remarks," *The Journal of Philosophy*, vol. 64 (1967), pp. 259–271.
[11] Hughes Leblanc, "That Positive Instances Are No Help," *The Journal of Philosophy*, vol. 60 (1963), pp. 453–462.
[12] Israel Scheffler, *Anatomy of Inquiry* (New York, A. A. Knopf, 1963).
[13] Brian Skyrms, *Choice and Chance* (Belmont, California, Dickenson, 1966)
[14] Judith Jarvis Thomson, "Grue," *The Journal of Philosophy*, vol. 63 (1966), pp. 289–309.
[15] —— "More Grue," *The Journal of Philosophy*, vol. 63 (1966), pp. 528–534.

VI
Goodman's Paradox
SIMON BLACKBURN

THE problem of justifying induction is a problem of showing why propositions of a certain sort are reasons for some other propositions. Two sorts of cases are naturally distinguished. The evidence considered may state that some of a kind of thing have a property ϕ, the proposition supported being that some or all others of that kind of thing are ϕ, or the evidence may concern just one thing, stating that it has been ϕ at certain times, the proposition supported being that the thing in question will be ϕ at some or all other times.

In terms of the first sort of case Nelson Goodman constructs his paradox by attempting to produce a predicate having the following three properties:

(1) The predicate is properly used to express a property θ which relates to a property ϕ which some of a class of things can be known to possess, in such a way that if those things are ϕ, then they are also θ.
(2) θ and ϕ are symmetrical with respect to confirmation, that is, if some of a class of things being ϕ is a reason for supposing other members of the class to be ϕ, then some of a class of things being θ is just as good a reason for supposing other members to be θ.
(3) The supposition that some unobserved things of a class are θ is inconsistent with the supposition that they are ϕ.

It is further clear that if Goodman can produce one such predicate, then for any ϕ he can produce any number of such predicates, giving any number of apparently equally reasonable but mutually inconsistent predictions about the properties of unobserved members of a class: this is paradoxical because it conflicts with the apparently necessarily true proposition that some such predictions are better supported by given evidence than others.

In terms of the second sort of case the paradox would be constructed by producing a property θ, relating to a property ϕ which a thing can be known to have possessed at given times, such that if it was ϕ at those times then it was θ at those times, θ and ϕ are similarly symmetrical with respect to confirmation, and the prediction that the thing will be θ at some given times is inconsistent with the prediction that it will be ϕ at those times.

In addition to setting up the paradox, however, Goodman also advances a solution to it. His solution is to deny that for the predicates he advances, condition (2) is satisfied, on the ground that these predicates are not so "well entrenched," that is, so often used, as others. This solution does not seem at all plausible, and it is the purpose of this paper to show that the arguments in favor of it are no good, and to show what is the correct solution.

Clearly the success of Goodman's endeavor depends upon his ability to produce such a paradoxical property, and upon his ability to establish his own solution. His own exposition[1] introduces the predicate as follows, t being a variable ranging over times:

> Now let me introduce another predicate less familiar than "green." It is the predicate "grue" and it applies to all things examined before t just in case they are green but to other things just in case they are blue. (P. 74.)

I think we can only properly take this introduction to mean that the predicate expresses what may otherwise be expressed by a truth-functional disjunction with the notion of being examined involved:

> *grue:* a thing is grue if and only if *either* the thing has been examined before t and is green, *or* the thing has not been so examined and is blue.

It is usual to let t be some arbitrary time in the future, say midnight on the last day of 1969, and I shall call this value T, and it is usual to take emeralds as the class of things about which we are concerned to predict the grueness or greenness. It is worthwhile pointing out that if we let t be some time in the past, no paradox arises. Some emeralds, namely those discovered and examined for color before whatever time it was are grue, those found since are not, unless, of course, they are blue. It is also clear that if t did take some value in the past, upon being presented with a green emerald I could *not* determine whether it is grue, for to *know* this I would have to know whether it was examined before the fatal date. The importance of this point will, I hope, emerge in due course.

There is no doubt that "grue" so defined satisfies the first and third conditions upon the paradoxical predicate. For consider all the emeralds which we have examined and know thereby to be green.

[1] Nelson Goodman, *Fact, Fiction, and Forecast* (London, Athlone Press, 1954).

Now we may be slightly unhappy about saying that we have therefore observed them to be grue, i.e., observed them to be examined before T and green, or not so examined and blue, but they certainly *are* grue, for they satisfy the first disjunct, and, having observed them and seen that they are green, we can properly claim to *know* that they are grue, given, of course, that we know that the time is before T. Consider also the prediction that some unexamined emeralds are grue: given that among these there are emeralds which come to be examined only after T, and therefore do not satisfy the first disjunct, we must conclude that they satisfy the second. That is, the prediction that emeralds not examined by T are grue is the prediction that they are blue, and is inconsistent with the prediction that they are green.

Any solution to the paradox with this predicate must therefore show that the second condition is not satisfied, that is, it must show that knowledge that some things of a class are grue is not a good reason for supposing all other things of that class to be grue, or for supposing that members of the class which are to be examined after a certain time will be grue. And indeed, it does appear plausible to suppose that such knowledge is not a good reason for such a prediction. For if we look at the definition of "grue," it certainly appears that such knowledge is not a reason for such a prediction because the property *differentiates* between examined and unexamined emeralds, so that to make such a prediction is to expect a difference between a class of emeralds including those which we have examined, and others, and to expect such a difference appears to be irrational. I believe that this reaction can be shown to be the right one, although a great deal must be said to defend it from certain objections.

Let us first consider a situation in which it is reasonable to suppose that emeralds examined after T will be grue. Because of a religious belief, an emerald-mining tribe always treats emeralds in a certain way before bringing them into the light of day; they believe that this averts evil spirits but do not believe that it alters the color of the gems. An inquiring tribesman knows that all emeralds are green by the time that they are examined, and he knows that the practice of treating the gems will cease at midnight on the last day of 1969: such is the nature of his religion. But he wonders what color emeralds are before being treated, being dissatisfied with the prevalent belief that the treatment does not alter their color, and to form an estimate he treats a sapphire in the same way, and the sapphire turns green, indeed many blue things turn green under the operation, and most non-blue things turn some other color. The tribesman now rationally believes:

(i) All emeralds examined before T are green.
(ii) If they had not been so examined, they would have been blue.
(iii) All emeralds are either observed before T and green, or not observed before T and blue.

In fact the tribesman is wrong about the color of undiscovered emeralds, but it seems to me that he has rationally come to believe that all emeralds are grue, and that the basis for this judgment is his belief in (i) and (ii). Now we might certainly believe (i), so we seem to differ from the tribesman in not believing (ii). A fair characterization of this difference is that he has reason to suppose that something occurs which *distinguishes* the color of those emeralds examined before T from that of those emeralds which are not, and we do not.

The story shows that it is possible to describe a situation in which coming to know that all hitherto examined emeralds are grue ought to increase one's confidence that all emeralds are grue. If the tribesman knows that things which are blue tend to turn green under the treatment, and that things which are green tend to turn some other color, and also knows that all examined emeralds will have been previously treated, and that the time is before T, then coming to know that all hitherto examined emeralds are grue ought to increase his confidence that all emeralds are grue, i.e., that the ones examined before T will be found to be green, and those not examined before T will be blue. Of course, this has nothing to do with the relative entrenchments of "green" and "grue" in the man's language: it is just that he is in the unusual, but not unknown, situation in which it is reasonable to believe that the conditions of observation affect the property observed.

The primitive objection to the use of examined things being grue as a reason for supposing all things of some sort to be grue is not just that grueness somehow involves a reference to a point in space or time; it is the more serious defect that it uses this reference to segregate emeralds including those which we have examined from those which we have not, but about which we are predicting, which is important. I am afraid that this is still very unclear, but I think it will become clearer when we have considered the argument which Goodman would use against this approach. For Goodman would certainly object that whether or not we think that a predicate does differentiate some members of a class from others, so that we cannot normally use some members of a class possessing the property it expresses as a reason for supposing the others to do so, is itself a matter of the entrenchment of the predicate within the language.

In other words, I think he would claim that nothing I could say about there being a difference between examined emeralds being grue and unexamined ones being grue could provide an alternative account to his, which depends upon the fact that we do not use the word "grue" very often.

Goodman's argument is very simple. He points out that we can define another new predicate "bleen" in exactly the same way as we defined "grue" but substituting "green" for "blue" and vice versa throughout. Thus a thing is bleen if and only if either the thing has been examined before t and is blue, or the thing has not been so examined and is green. It is then possible to "define" blueness and greenness in terms of grueness and bleenness:

> *blue:* a thing is blue if and only if *either* the thing has been examined before t and is bleen, *or* has not been examined before t and is grue.

Similarly for "green" with "bleen" and "grue" reversed. I think it can be seen that with these two predicates in the uses described for them, this does give a "definition" of blueness. For the disjuncts expand upon substituting the original definitions of "grue" and "bleen" to give: "a thing is blue if and only if *either* the thing has been examined before t and (has been examined before t and is blue or has not been so examined and is green), *or* the thing has not been examined before t and (has been examined before t and is green or has not been so examined and is blue)." And it is true that a thing is blue if and only if this is satisfied. I mention this because it is not obvious that there is this interdefinability, and indeed with some predicates which people have taken Goodman to be talking about, it does not obtain.

Having established interdefinability of "grue" and "bleen" with "blue" and "green" Goodman continues:

> But equally truly, if we start with "grue" and "bleen" then "blue" and "green" will be explained in terms of "grue" and "bleen" and a temporal term.... Thus qualitativeness is an entirely relative matter and does not by itself establish any dichotomy of predicates. This relativity seems to be completely overlooked by those who contend that the qualitative character of a predicate is a criterion for its good behavior. (P. 79.)

And I think there is no doubt that he would use the same argument against the attempt to show that his paradoxical predicates really do differentiate between examined and unexamined emeralds. But is this argument really sufficient to show that qualitativeness, or differentiation, is an entirely relative matter?

The interdefinability does show that *if* we are familiar with the uses of "grue" and "bleen" we can *describe* the uses of "blue" and "green" in terms of them and conjunction, disjunction, and a temporal term, and "examined." It does not show that this could be an *explanation* of the uses of "blue" and "green," because it does *not* show that we could be familiar with the uses of the paradoxical predicates without being familiar with the uses of "blue" and "green" to begin with. However, there is a more important point. Goodman might claim that somebody could use "grue" and "bleen" correctly without being able to use "blue" or "green" or any synonyms, and even although the interdefinability does not show this, it would be difficult to establish that it is impossible. But even if he held this, an asymmetry remains to which the interdefinability is completely irrelevant. This is that you could not come to know that a thing is grue without *either* examining it *and* knowing whether it is before or after midnight on the last day of 1969, *or* knowing whether or not it was examined before that time. Suppose for example that I am examining emeralds. I cannot tell, just by looking at them, whether they are grue; because I cannot tell, just by looking at them, what time it is or at what time they were first examined. This epistemological point is perhaps clearer if we consider someone presented with a tray of green emeralds at some time after the end of 1969: he cannot know which ones are grue without knowing which ones were found and examined before the end of 1969, and therefore satisfy the first disjunct in terms of which "grue" is defined.

I pointed out above that the interdefinability does not by itself show that there could be a language in which "grue" and "bleen," in their described uses, are primitive, with the uses of "blue" and "green" *explained* in terms of them. But even claiming that there could be such a language is not enough to show that whether or not a predicate in a certain use differentiates examined from unexamined objects is a language-relative matter. For in pointing out that knowing whether a thing is grue involves knowing facts about the time at which it was first examined I am not contradicted by someone stating that "grue" could be a *primitive* predicate in its described use, but only by someone claiming that it could be an *observation* predicate in its present use, where an observation predicate is at least one whose application does not entail knowledge of the time, or of whether or not the thing to which it is applied was examined before a certain time. I think it should be fairly clear that no facts about interdefinability have the slightest relevance to showing that "grue" and "bleen" could be observation predicates in this sense.

I am afraid that I still have not put this point in the clearest possible way, although I do not know if I can make it any clearer. I am in effect claiming that there is an epistemological asymmetry between grueness and similar properties on the one hand and blueness and similar properties on the other. The epistemological asymmetry is that to know that something is grue entails knowing not only what it looks like, but also what time it is, or at what time it was first examined, whereas knowing that something is blue does not entail knowing either of these things. Furthermore, no facts about how grueness or blueness expressed in a particular language can alter this asymmetry; nor can facts about the possibility of people having extraordinary sensory faculties do so. The asymmetry upon which the solution of the paradox depends is *not* a psychological one, it is not one which depends upon any contingent fact about the way in which we, with our sensory abilities, find it natural to classify things. I am not, of course, denying that there could be a people who can tell immediately, at any time, whether something with which they are presented is grue. We cannot do this, but we can consistently imagine someone, perhaps with some extraordinary sensory faculty, who could at any time state correctly whether a thing is grue or not, and apparently just by looking at it. What I am denying is that anyone could do this without equally being able to tell, just by looking at the thing, whether or not it was examined before the end of 1969, or whether or not the time is before or after the end of 1969.

We can, for example, imagine somebody who appears to use the word "grue" in the way described, and who is mining emeralds throughout New Year's Eve, 1969. He correctly says, as he turns up green ones and examines them, "Ah, these are grue," until at midnight he turns up a green one and says, "Funny, here is one which is not grue." But now suppose that we ask him the time, and he says that he doesn't know if it has passed midnight, and we ask him if anyone had ever previously dug up that emerald, examined it for color, and replaced it in the ground, and he says he doesn't know that either. We have, it seems to me, *conclusive* evidence that he didn't know that the emerald was not grue, so that either he was not using the word "grue" in accordance with the description we gave of its meaning, or he was mistakenly claiming knowledge.

Whether or not examined objects of a class possessing a property differ from unexamined ones possessing that property is determined by other conditions than the length of the predicate expressing that property, or the overtness of occurrence of words like "examined" or "midnight on the last day of 1969" in that predicate. These other conditions refer to the use of the predicate, in particular to what we must know to know that what it expresses is true of a given object. These epistemological conditions are obviously not functions of the linguistic form of the predicate ("grue" is as short as "blue"), nor of the length of time for which it has been used. Take a parallel case. There could, I imagine, be a language in which the words "squabble," "squot," and "blot" are primitive, with the following uses:

squabble: a thing is squabble iff it is both square and blue.
squot: a thing is squot iff it is square and not blue.
blot: a thing is blot iff it is blue and not square.

We can fairly easily imagine children being taught the use, or at least the application, of these words, before knowing any synonyms of "square" and "blue"; they would later learn that a thing is square iff it is squabble or squot, and blue iff it is squabble or blot. But this does not show that these children could recognize that something is squabble without recognizing that it is square or without knowing its color; it is simply irrelevant to the logical fact that you cannot know that something is square and blue without knowing that it is square.

This conclusively refutes the supposition that interdefinability shows symmetry in everything but frequency of occurrence; in particular it refutes the idea that whether or not a predicate expresses something differentiating between examined and unexamined members of a class is a matter of its frequency of occurrence, and *must* be a matter of something like that because of interdefinability. However, I expect it will be felt that there remains some point to the paradox, and that I have not yet presented a solution of it. I have so far claimed that grueness differentiates between examined and unexamined emeralds in a way in which greenness doesn't, that this is shown by considering what it is to know that a thing is grue, and that these epistemological asymmetries are not simply language-relative. I can say nothing more about the last point, but I shall try to explain the first a little more, and explain in what way it leads to a solution of the paradox.

Suppose someone said: "It is all very well claiming that the prediction that all emeralds are grue differentiates those emeralds examined before a certain time from the rest, in respect of color. Why should this matter? After all, the prediction that all emeralds are green differentiates those examined before a certain time from the rest, in respect of grueness. So this description does nothing to distinguish one prediction from the other in irrationality, given the evidence from examined emeralds." We must, I think, only remember the epistemological asymmetries to realize that there are differentiations and differentiations, and that it might be quite possible to separate them in point of irrationality. After all, a man who treats all men equally well segregates those whom he has seen, and (seen and treated well, or not seen and treated badly) from those whom he has not seen, and (seen and treated badly, or not seen and treated well), but he is not a segregationist for all that. If we call the expression in the first bracket A and the expression in the second B, then he is not exercising prejudice in treating men he has seen in way A and men he has not in way B, for to do this is to treat all men equally, however often we use the expressions A and B or any shorter synonymous expressions.

Similarly with grueness. To know whether a thing is grue entails knowing whether it was examined before T, and had one feature, or was not, and had another. To know whether a thing is green does not entail knowing whether it was examined at any time at all. So differentiating things examined before T from others in point of grueness need not be differentiating at all, whereas doing the same in respect of greenness must be.

It may of course still be asked why it is irrational to expect a difference between things examined before a certain time and others, but here the spurious nature of the "new riddle of induction" is revealed. For once we have established that Goodman's predicate really does distinguish between those of a class including those we have observed, and the rest of that class, it is apparent that nothing more needs to be done to establish the irrationality of believing that it applies to all members of a class, than needs to be done to establish the rationality of induction. That is, suppose that I know something non-differentiating (by the epistemological criteria) to be true of all examined members of a class of things. You form the corresponding Goodman predicate, parallel to "grue," for some value of t. I can establish that you are differentiating between instances examined before T and those which are not, in a way in which I am not. But then, in believing that all members of the class have the Goodman property you are simply postulating a difference between some of the things, including those examined, and others. If, unlike the member of the emerald-mining tribe which I considered, you have no reason for this difference, your prediction is irrational, but its irrationality simply follows from its being rational to expect the future to be like the past. But establishing this is the old riddle of induction, not a new one. In other words, Goodman makes it look as though there is *no* problem of justifying taking the future to be like the past, but there *is* a problem of choosing between the different respects (grueness, greenness) in which the future may be like the past. However, this is not correct, because this new problem reduces to the old problem.

I think this disposes of Goodman's paradox in the form in which he presents it, and disposes of it without relying upon the frequency of occurrence of different predicates in a given language. I do not claim to have shown that the second condition upon the paradoxical predicate is *not* satisfied, but I do claim to have shown that to demonstrate that it is satisfied is no other problem than demonstrating the rationality of induction. That is, suppose that I wonder why, in general, observed things of a type all having a property is a reason for unobserved things of that type having it, and you point out the existence of a property, or indeed of many properties, which the observed things possess, but such that, if the unobserved things possess them, then they differ from the observed things. I need do nothing more to show that it is irrational to expect the unobserved things to possess these properties than I need do to show that it is irrational to expect the unobserved things to differ from the observed: but this is the rationality of taking the observed as a guide to the unobserved.

There are, however, other forms of Goodman's paradox than the one which he presents. People have taken him to be defining a different predicate from the one which I described, and it may seem possible that some other property gives a more difficult problem. So for completeness it is necessary to say something about these other candidates. Unfortunately it is seldom very clear just what the other candidates are supposed to be, and just what the predicates are supposed to mean. For example Kyburg says that "grue" is to mean "green and occurring before 2000 A.D. or blue and later than that."[2] But it is very hard to see just what this means. It is quite clear whether an event, or a thought, occurs before 2000 A.D., but hardly so clear whether an object of the sort that can be green does, or indeed what sense can be given to saying that an object occurs at a certain time. When, for example, did my piano occur? If an emerald came into existence many millions of years ago, and is mined in the year 2001 A.D., when did it occur? Is the proposition that an object is grue, in this sense, supposed to entail that it changes color in the year 2000 A.D.? I am afraid that I cannot see the answer to any of

[2] Henry E. Kyburg, Jr., "Recent Work in Inductive Logic," *American Philosophical Quarterly*, vol. 1 (1964), p. 263.

these questions, and it is difficult to see what Kyburg supposes his predicate to mean, in the absence of these answers.

Hacking[3] gives a different example. He says:

> Let "blight" mean "black until the end of 1984 and white thereafter"; let "wack" mean "white until the end of 1984 and black thereafter." (P. 41.)

But the trouble with this is that it does not satisfy the first condition on the paradoxical predicate. For clearly a thing can be black without being blight, and wack without being white. To be blight an object must satisfy a conjunctive condition; black until the end of 1984 *and* white thereafter. To know of a particular thing that it is blight, I must therefore know that it is black now, will continue to be so until the end of 1984, and will then turn white and remain so until it ceases to exist. I certainly do not know this to be true of any object, even although many objects are known by me to be black. Of course, the hypothesis that a thing is blight is consistent with the hypothesis that it is black now, and indeed the hypothesis that an object is blight entails that it is black before the end of 1984. But observation that a thing is, at this time, black, is *not* observation that it is, at this time, blight. To be, at this time, blight it must be true, at this time, of the object, that it is black before the end of 1984 and white thereafter. And this cannot be true, now, of an object without that object being white after the end of 1984, any more than it can be true, now, that there will be a battle tomorrow, without there being a battle tomorrow. In other words, Hacking's claim that "blight" and "wack" provide a problem about projectibility is simply false, for we have no knowledge that any particular thing is blight, or even that any particular thing is blight at this time, although we know that many things are black at this time.

Hacking's misdescribed predicate seems to be an attempt to find a paradox involving predictions about the future color of some one object, rather than predictions of the color of other objects of a particular class, as Goodman's original presentation did. Barker and Achinstein[4] define the predicate as follows:

> it applies to a thing at a given time iff either the thing is green and the time is prior to t, or the thing is blue and the time is not prior to t. (P. 511.)

[3] Ian Hacking, *The Logic of Statistical Inference* (Cambridge, University Press, 1965).
[4] S. F. Barker and Peter Achinstein, "On the New Riddle of Induction," *The Philosophical Review*, vol. 69 (1960), pp. 511–522.

It will be seen that this definition makes no use of the notion of "being examined," so whatever predicate Barker and Achinstein have in mind, it will be different to the one which Goodman seemed to be considering. But there are certain things which Barker and Achinstein must not be taken to have in mind if they are to produce a paradoxical property. In the first place they must not be considering a word which changes its meaning. Consider the word "grue*" which now means exactly the same as "green," but, owing to the dictates of some Academy, is to change its meaning at the end of 1969 to become synonymous with "blue." There would be difficulties about implementing such a decree, but I do not see that it is logically or practically impossible to obey. It would so far as I can see be quite correct to say that a thing is (called) grue* iff either the thing is green and the time is prior to T, or the thing is blue, and the time is not prior to T. Thus "grue*" satisfies the definition given. But no paradox can arise with this word. For the word now means just the same as the word "green": there is no question of my expressing inconsistent predictions by saying that something will at any time be grue* and green. The hypothesis that anything will be grue* is the hypothesis that it will be green, and observation that a thing is grue* is observation that it is green. After the end of 1969 it will no longer be correct to express such hypotheses and observations by use of the word "grue*," but this does not mean that predictions made now by using it will be falsified unless the things concerned turn blue. To suppose this would be to confuse "This thing will be grue* in 1970" with "The word 'grue*' will apply to this thing in 1970." But these are obviously different: the latter but not the former entails a belief about what a word will be used to do at a certain time, and knowing the decree of the Academy I may well wish to say the former but not the latter of some green thing, and the latter but not the former of some blue thing. "Grue*" is not a word which has a different use or meaning from an English word, and hence no paradox arises.

Also, it will not do to simply consider Barker and Achinstein as defining a disjunction: either the thing is green before time t, or the thing is blue after time t. No paradox could arise with a predicate meaning this, for it is just a disjunction weaker than the property of being green before a certain time. An emerald which is green now would be grue in this sense, and would be grue at any time, whether or not it changed color after the favored value of t. Similarly, in this sense stars and the sea are grue, for they will be blue after any likely value of t.

The correct expression of a paradoxical property which Barker and Achinstein might have in mind is the following:

At any time t, a thing x is GRUE at t if and only if:
$(t<T \supset x$ is green$)$ and $(t>T \supset x$ is blue$)$.

A thing is omnitemporally GRUE iff $(t)x$ is GRUE at t. It has been pointed out to me by J. J. Altham that this is logically equivalent to a truth functional disjunction formally similar to Goodman's own predicate:

At any time t, a thing x is GRUE at t if and only if:
$(t<T$ & x is green$) \vee (t>T$ & x is blue$)$.

This differs from Hacking's property in that we can observe something to be GRUE at a particular time just by knowing that the time is less than T, and observing that it is green at the time or knowing that the time is after T, and seeing that it is blue at the time. So we can certainly come to know, of a particular thing, that it is GRUE at a particular time. Furthermore, the proposition that it is omnitemporally GRUE (assuming that it exists after T), or the proposition that it is GRUE after T, is inconsistent with the proposition that it is omnitemporally green, or green after T. For it will be GRUE after T if and only if it is blue after T. So with GRUEness the first and third conditions upon the paradoxical property, relating now to predictions about one particular thing, are satisfied, and as with Goodman's own property, any solution must deny that a thing's being GRUE at certain times is a reason for supposing it to be GRUE after T. Also GRUEness makes no reference to the notion of "being examined," so some of the things which I said in discussing Goodman's own predicate will not be straightforwardly true of it.

But so far as I can see, the general form of the solution is exactly the same: GRUEness really does differentiate between the property of things in virtue of which they are GRUE at some times, which include all the times at which we have examined them, and the property in virtue of which they are GRUE at others. This differentiation is not just a matter of entrenchment, for there is the epistemological asymmetry between GRUE and green, namely that to know whether something at which I am looking is GRUE at the time, I have to know whether it is before or after T. If, for example, I believe that an emerald is omnitemporally GRUE and I am looking at it on New Year's Eve, 1969, and it really does turn blue, I shall be able to tell when it is midnight, and if someone disputes whether it was midnight when it changed color, I might have to renounce my belief that it is omnitemporally GRUE, or telephone Greenwich or somewhere to find out the exact time. And again, the irrationality of differentiating between the properties of an object at some times, including all those at which we have observed it, and others, about which we are concerned to predict, is nothing other than the rationality of taking what has been observed as a guide to what will happen.

Perhaps this is clearer if we consider a different example. Goodman's paradox is a logical paradox, and is supposed to have equal force when we consider any prediction, not just predictions of color. Suppose then that we take the predicate "is a wubble," defined as follows:

At any time t a thing x is a wubble at t if and only if:
$(t < T \supset x$ is a wall) and $(t > T \supset x$ is rubble).

It is, I should hope, perfectly obvious that to know whether a thing is now a wubble I have to know whether the time is now before or after T, and equally obvious that, looking at a wall which does in fact collapse into rubble, I cannot tell whether it was omnitemporally a wubble unless I know that the time at which it collapsed was T. Nor is this a psychological fact: it is logically impossible that these things should be known without the facts about the time being known. As in the case of the miner cited previously, if someone claimed to know that a thing is a wubble, by looking at it, then he is refuted if it is shown that he does not know the time to be before or after T. If someone claims to know that a thing was a wubble throughout a certain period, he is conclusively refuted if it is shown that he does not know whether the time passed T during that period.

One interesting and puzzling feature of Goodman's argument is its relation to some passages in Wittgenstein.[5] Apart from their other interest these passages might be taken to raise the following sort of scepticism: might not somebody in fact mean something different by a predicate, even although everything we could get him to say at present would be just what would be expected if he understood it to mean what we mean by it? But I think it should be clear now that this interesting and difficult problem has nothing to gain from consideration of Goodman's predicates. For although the cases have this in common, that somebody might misunderstand explanations of the use of "green," and take it to mean GRUE, and although his mistake might not be detected, since he always applies the word to the right things until New Year's Day 1970, nevertheless his use of "green" would be different before that time, and would be detectably so. The difference is detectable both in what

[5] Particularly *Philosophical Investigations* (Oxford, Basil Blackwell, 1953), § 185; *Remarks on the Foundations of Mathematics* (Oxford, Basil Blackwell, 1956), I, 3.

he would say has to be known to know that a thing is green, and what he would say about the consequences of something being green in 1970. Again this latter point is more evident if we consider properties having more consequences than colors: a man who thinks that "wall" means what we have defined "wubble" to mean will think that consequences of a thing being a wall in 1970 are that you can't park your car under it, grow plants up it, build houses against it, and so on.

In this paper I have tried to give a more careful statement of Goodman's paradox than is usual, and tried to show that a satisfactory criterion for distinguishing his properties from normal ones can be given. The existence of this criterion is not affected by the possibility of intertranslation, nor the possibility of people being able to perceive other properties of things than we can. This conclusion shows that when discussing induction we must restrict consideration to those properties which do not use some arbitrary feature (time of occurrence, spatial position, or anything else) to differentiate between known and unknown instances. It also shows that whether a property does this is not contingent upon facts about our abilities and language.

I said that there is nothing more to the "new riddle of induction" than there is to the classical problem of why to expect what has not been observed to be like what has been observed. This is not to say, however, that there do not exist further problems of comparison of projectibility of different hypotheses which are not solved with the solution of Goodman's paradox, and which do not reduce to the classical problem. In the problem of which function is best supported as describing the co-variation of a number of physical parameters, given their values at various points, the techniques I have used for showing that one hypothesis expects a difference where the other does not, will not be straightforwardly applicable. All I claim to have done is reduce this problem to the classical one for one special way of fabricating conflicting hypotheses. It is finally worth remarking that Goodman puts the old riddle in a form which does enable us to see that some approaches will not solve it. In particular any approach which is unable to distinguish the hypothesis that a thing is omnitemporally GRUE from one that it is omnitemporally green, because of their equal falsifiability, or equal syntactical simplicity, or because they both entail the observations so far and are indistinguishable in terms of prior probabilities, is not going to be adequate to solve the old problem.

Churchill College, Cambridge

VII
Assuming, Ascertaining, and Inductive Probability

STEPHEN SPIELMAN

1. The Aim of the Paper

TYPICAL situations in which a decision-maker might want the counsel of a method for choosing a course of action on the basis of limited evidence are those in which he must adopt one of several specified courses of action,[1] the outcomes of which, if adopted, depend on the truth-value of an hypothesis[2] h (or on the truth-values of the members of a set of hypotheses) for which he possesses a relevant, inconclusive observation report e assumed by him to be true. It is generally held that any decision-making method for such situations should be based on the rule that if e represents all the decision-maker's relevant observational data, and if it makes sense to talk about the probability of h given e (i.e., given that e is true), then this probability should, in conjunction with an evaluation of the risks involved, be a guide to the decision-maker's choice. Precise formulations of this rule, which will henceforth be referred to as Rule C, usually involve some form of the principle of maximizing expected utility.[3]

Since any probability which ought to be a guide to our inductive or "betting" behavior[4] can appropriately be labelled an "inductive" probability, Rule C states, in effect, that for situations of the sort described above, the probability of h given e is the inductive prob-

---

[1] E.g., to decide whether or not to declare a drug safe, to accept an hypothesis, or to declare a man guilty of a crime.
[2] "Hypothesis" will be used as a label for any proposition whose truth-value is not known by a decision-maker and which is of interest to him.
[3] Cf. Rudolph Carnap, "The Aim of Inductive Logic" in *Logic, Methodology and the Philosophy of Science*, ed. by E. Nagel, P. Suppes, A. Tarski (Stanford, California, Stanford University Press, 1962), pp. 303–318. Also L. J. Savage, *The Foundations of Statistics* (New York, Wiley & Sons, 1954).
[4] "Inductive behavior" will be used in the sense of Jerzy Neyman who defines it as "... the adjustment of our behavior to limited amounts of information," *First Course in Probability and Statistics* (New York; Holt, Rhinehart & Winston, 1950), p. 1.

ability of h for the decision-maker, provided that e represents his total relevant (to h) observational data.

I intend to show by what follows (1) that Rule C is not correct under its customary interpretations because it takes the wrong probabilities as guides to action, and (2) that since Rule C is the basic presupposition of the Carnapian *approach* to inductive logic, all applications of this approach are wrong *in principle*.

In what follows it will be assumed that propositions are bearers of probability and that the propositions under consideration in a problem belong to a set of propositions for which there exists a unique numerical measure of probability. These assumptions are not essential to the discussion which follows; if the arguments hold under them, they hold for any adequate interpretation of probability, the obvious changes being made.

2. Some Preliminary Considerations

Before presenting the examples on which my case rests, some fairly obvious and pedestrian points about the proper interpretation of statements of the form "The probability of p, given q, is equal to a" must be made. Following customary notation, the measure of the (conditional) probability of p, given q, will be represented by "$P(p/q)$" and the measure of the "unconditional" probability of p will be represented by "$P(p)$." The functor P in contexts of the form "$P(p/q)$" represents, of course, a different function than it does in contexts of the form "$P(p)$." It will be assumed that instances of both kinds of functions satisfy the usual conventions of probability measurement.

Statements of the form "$P(p/q) = a$" cannot be interpreted as asserting that if q is true, $P(p) = a$. For, under such an interpretation either $P(p) = 0$, or $P(p) = 1$ for any p whatsoever, which is clearly absurd. This would follow from the fact that $P(p/p) = 1$ and $P(p/\text{not-}p) = 0$.

Nor can $P(p/q)$ be interpreted as $P(q \supset p)$, for $P(p \supset \text{not-}p) = P(\text{not-}p)$, (since not-$p$ is equivalent to $p \supset \text{not-}p$) and this latter value may not equal zero whereas $P(\text{not-}p/p) = 0$.[5] It is also false that $P(p/q) = P(\text{if } q \text{ then } p)$ where "if ... then - - -" is interpreted as asserting the existence of some sort of physical relation (e.g., casual or temporal) between the states of affairs described by p and q. The argument just presented doesn't hold for this case because "if p then

---

[5] However, it should be noted that $P(p/q) \geq P(r/q)$ if and only if $P(q \supset p) \geq P(q \supset r)$.

not-p" is not equivalent to "not-p." But consider the sentence "If on the next roll of the dice the red die results in 4, the roll of the green die will result in 2." Letting a abbreviate the antecedent and c the consequent, under the assumption that the outcomes of rolls of different dice are independent, $P(c/a)=P(c)$, whereas we certainly would not want to say that $P(\text{if } a \text{ then } c)=P(c)$.[6]

The proper interpretation of the formula "$P(p/q)=a$" seems to be this: if the truth of q is *assumed*, then the new probability of p is to be measured by a. This may be expressed in symbols by "if q is assumed, then $P'(p)=a$," where P' represents the new measure of probability that results from assuming q.

The basis of this interpretation is the fact that probabilities are "corpus relative." By this I mean that it does not make sense in a practical case to ask what the probability of a proposition (or event, event-type, class, etc.) is, without specifying, at least implicitly, a corpus consisting of a problem framework and some assumptions and/or a body of data *on the basis of which* the question may be answered. For example, it doesn't make sense to ask what the probability of a person's having blood type O is, with the purpose of applying the answer to particular cases, without either indicating or assuming something about the population the probability-value is to be applied to and the method by which the individuals to which it is to be applied have been or will be selected.[7]

To ask what the measure of the probability of p given q is, is to ask the following question: If P is the measure of the probabilities of the propositions I am currently concerned with, how must I revise P in order to measure the new probability of p that would result were I to include q in my probability corpus as an assumption? The generally accepted answer (although for different reasons) to questions of this sort is that if P is defined over q (as is usually the case), the revised measure—the measure of the conditional probability of p given q—is $P(p \,\&\, q)$ divided by $P(q)$, in symbols, $P(p \,\&\, q)/P(q)$.

This answer will be accepted as correct in what follows. The burden of the examples presented below is to show that it does not also provide, as is commonly believed, an answer to questions of the form "Were I to *ascertain* that q is the case, how should I revise the measure of the probabilities of the propositions I am presently concerned with?"

---

[6] If '⊃' is replaced by "if ... then - - -" the fact noted in the above footnote fails to hold.

[7] It does not follow that expressions of the form "$P(p)=a$" are meaningless. They have a perfectly legitimate use in contexts where there is no need to explicitly formulate the corpus to which P relates.

K

3. THE DICE EXAMPLE[8]

Imagine that you and I are betting on the rolls of a pair of dice—one red and the other green, and that the roller of the dice is at the other end of the room so that we cannot see the results of the rolls, although we are willing to assume that his reports about them are accurate. Let us also assume that the dice are perfectly "symmetrical."[9]

The dice are rolled before we bet. Rule C recommends that in the absence of any information about the results of a roll other than that the dice were rolled we take the fair betting odds against sevens to be 5 to 1, since the probability of sevens for such a case is 6/36 or 1/6 under the assumption of symmetry. To say that fair odds are 5 to 1 is to say that if I want to bet on sevens I should accept odds no lower than 5 to 1, and that you should not offer odds in excess of 5 to 1, provided that the amounts of money you and I are to bet are small in relation to our incomes (more accurately, if our utility for money is approximately linear for the range in our incomes covered by the stakes).[10] Surely the odds that Rule C recommends are appropriate for this case.

Suppose that the roller informs us that at least one three has been rolled. Call the sentence expressing this observation report e and the hypothesis that sevens was rolled h. If Rule C is correct the probability of h given e —$P(h/e)$— is the inductive probability of h for us and should determine the fair odds against sevens for the present case. $P(h/e) = P(h \& e)/P(e) = (2/36)/(11/36) = 2/11$. Thus fair betting odds against sevens are 9 to 2 if Rule C is correct.

But should you be willing to offer these odds? Does the information proffered by the roller really make it more likely that sevens was

---

[8] A simple, abstract version of this example was presented during a course in mathematical probability I attended several years ago as a cute little example of how one cannot trust one's intuitions in probability and should calculate probabilities using theorems in a slavish way. After an illuminating discussion with Robert Seid, I began to see why the instructor's interpretation completely missed the point of the example.

[9] "Symmetrical" may here be taken in an epistemic or relative frequency sense.

[10] Fair odds are computed from inductive probabilities in the following manner: Let h be an hypothesis whose truth-value you and I are betting on and let $P(h)$ be the inductive probability of h for both of us. Suppose that I will give you \$$s$ if h turns out to be false and you will give me \$$t$ if it is true. Then t to s are the odds against h that we have agreed upon. They are "fair" if and only if the expected value of the bet is non-negative for both of us, which is the case if and only if my expected value is zero. My expected value is $P(h) \times t - P(\text{not-}h) \times s$, which is equal to zero if and only if $t/s = (1-P(h))/P(h)$. In general, fair odds against an hypothesis h are $(1-P(h))/P(h)$ provided that $P(h)$ is the inductive probability of h and the value of the stakes are "small."

ASSUMING, ASCERTAINING, AND INDUCTIVE PROBABILITY 147

rolled and hence justify odds lower than for the case of no report? I think not. First of all, note that for $x = 1, 2, 3, \ldots, 6$ we shall have $P(h/\text{at least one } x \text{ was rolled}) = 2/11$. Now suppose that just as the roller had started to utter the word "three" in "At least one three has been rolled" a truck outside blew its horn so loudly that we couldn't hear this word, although we could make out the rest of the sentence. Thus all we would know is that at least one – – – has been rolled. Would there be any point in asking the roller to repeat the word we didn't hear, so far as betting on sevens is concerned? Would we gain any relevant information? Not on the basis of Rule C. For we know that no matter what number-word he uttered, the value of the probability which would guide our bets would be the same, namely 2/11. Thus no relevant information with respect to the hypotheses was lost to us by our failure to hear the word "three," according to Rule C. But to know that some instance of "at least one x has been rolled" is true, is not to have any information over and above the information that the dice have been rolled. Hence fair betting odds against sevens should be the same as for the case in which the roller offers no information other than that the dice were rolled, namely, 5 to 1. This conclusion was reached on the assumption that Rule C is correct, i.e., that the conditional probabilities under scrutiny would be proper guides to betting behavior were we to learn that the conditioning event occurred. Thus it seems that Rule C leads to different fair betting odds for one and the same situation, and in this sense can be said to lead to inconsistencies. The argument presented above also seems to show that $P(h/e)$ is not the inductive probability of h even though e represents all our available observational data; it suggests that the inductive probability of h is 1/6.

These points are too important to rest on a single argument. Consider the following: If the roller had told us after rolling the dice that three is uppermost on the *red* die, fair betting odds against sevens should have been set at 5 to 1 according to Rule C, since $P(h/3 \text{ on red}) = P(h/3 \text{ on green}) = (1/36)/(6/36) = 1/6$. Suppose that the noisy truck outside had blown its horn just as the roller uttered "red" so that we couldn't hear this word. Then we would know that a certain unspecified die has 3 on its uppermost face. There would be no point, however, in asking the roller which die he mentioned because the probability is 1/6 regardless of which die it is. In other words, by so asking we would not gain any information which would make a difference so far as betting on sevens is concerned—according to Rule C. But knowing that a certain unspecified die has

3 on its uppermost face is equivalent to knowing that at least one 3 was rolled. Thus on the basis of our rule, being told that at least one 3 was rolled is equivalent, so far as betting on sevens is concerned, to being told that the number on the red die is 3. But according to this rule fair odds for the latter case are 5 to 1, whereas for the former case they are 9 to 2. Again, the rule seems to lead to an inconsistency.

At this point you may suspect that the conclusions presented above are based solely on the peculiarities of a special and unrepresentative example, and are not instances of a fairly deep fact about inductive probability. Such suspicion will be more difficult to sustain upon consideration of the fact that the probability of sixes given that at least one x has been rolled $= 2/11$ for $x = 1, 2, 4,$ and 5; $1/11$ for $x = 3$; and 0 for $x = 6$. It would clearly be ridiculous to offer 9 to 2 odds against sixes if the roller informs you that at least one x has been rolled for $x = 1, 2, 4,$ and 5; and 10 to 1 for the case $x = 3$. There seems to be no difference between these cases which would justify the different odds. By an argument parallel to the second noisy truck argument one can easily show that these are not the fair odds. Such an argument would suggest that correct odds are 5 to 1 for all five cases (of course the probability for $x = 6$ is appropriate).

4. The Card Example[11]

Imagine that you and I are playing two-card poker with a deck consisting of eight cards—four aces (A) and four kings (K) of four different suits: hearts (H), spades (S), clubs (C), and diamonds (D). Assume that before each deal the cards are "perfectly" shuffled.

Suppose that you are very suspicious of me so that when you open your cards you hold them so close to your body that my spy with the high powered telescope can just barely see one of them. Suppose also that when you get your cards you rapidly shuffle them face down so that my spy can't tell if the card he can see is the first or the second one dealt.

The cards are dealt. The spy informs me via the miniature radio receiver placed in my left molar that you have an ace of hearts. I have a king of hearts and a king of diamonds and wish to know what fair betting odds against the hypothesis that you have two aces are. There are 56 equiprobable (due to "perfect" shuffling) hands that a player can receive, taking into account the possible order

[11] I am indebted to Charles Pailthorp for bringing to my attention a problem in Irving Copi's *Introduction to Logic* which suggested this example.

with which the cards were dealt. Since I have the *K* of *H* and the *K* of *D* there are only 10 possible hands that you might have. Since six of these involve your having two aces, the probability of your having two aces given the observational data is 3/5. According to Rule C, I should bet accordingly. Note that this 3/5 figure applies regardless of the suit of the ace my spy saw in your hand.

Now suppose that I have been winning a lot and you have become so suspicious of me that my spy can see only a tiny corner of one card—enough to see its value but not its suit. I have as before a *K* of *H* and a *K* of *D* in my hand but this time the spy's observation report contains the information that you have an ace, suit unknown. There are 28 possible hands that you might have and 12 of these correspond to a pair of aces, hence the probability of the hypothesis that you have a pair of aces given the observational data $= 12/28 = 3/7$.

Should this probability be a guide to my betting behavior? That is, are fair odds against your having two aces 4 to 3? Not if fair odds for the previous case are 2 to 3. For not knowing the suit of the ace my spy saw should make no difference at all: Since the probability of two aces given that I have a *K* of *D* and a *K* of *H* and you have an *A* of *x* is the same for the four possible values of *x*, it shouldn't matter whether or not I know the suit—according to Rule C.

The two noisy truck arguments presented above have parallels for this case. The first one would correspond to the case in which the sound of the horn made it impossible to hear what the suit of the ace the spy saw was, and the second would correspond to the case in which the spy couldn't determine the suit of the ace but could tell if he was looking at the first or at the second card dealt. I will spare the reader the details. They would converge to the same conclusion: Rule C leads to two different fair betting odds against the hypothesis that you have two aces if I possess an observation report to the effect that you have an ace, suit unknown, and I have a *K* of *H* and a *K* of *D*. The arguments also *suggest* that fair odds for this case are the same as for the case in which the suit of your ace is known, namely, 2 to 3.

5. A POSSIBLE OBJECTION

The most plausible initial objection to the claim that the examples discussed above show that something is radically wrong with Rule C is that the "observation" reports for which the rule has broken down are somewhat queer and atypical, and that the rule does not break down for the cases in which they are not so queer. On this objection they are queer because they really are equivalent to *disjunctions* of what may be called "direct" observation reports, whereas in typical cases observation reports are equivalent to *conjunctions* of "direct" observation reports. For example, "At least one three has been rolled" is equivalent to "Either the red die has 3 on its uppermost face or the green die has 3 on its uppermost face." An objection might run: "The rule works for each of the two disjuncts—which clearly express what the roller actually sees—but it does not work for their disjunction." This objection does not get to the heart of the matter. First of all, in the card example the spy's report that my opponent has at least one ace seems to be, under the condition of the example, every bit a "direct" report of what the spy sees as the report that my opponent has an ace of hearts (for which the rule doesn't *seem* to break down). Secondly, many standard types of observation reports are equivalent to disjunctions of "direct" observation reports, e.g., reports of the form "$z\%$ of the 100 people interviewed fall in category F." Thirdly, it can be shown that the examples discussed above are far more complex than has been indicated so far and that the rule doesn't hold for them even for the cases of "direct," "conjunctive" observation reports (cf. Sect. 7).

6. A Carnapian Analysis of the Examples

In this section it will be shown that the misleading (for inductive behavior) probability-values obtained above are the ones that Carnap's system recommends as guides to betting behavior. The discussion will be based on Carnap's presentation in the Schilpp volume on his work[12]—which reflects his most recent views on inductive logic in print.

Consider the dice example. There are three plausible ways of approaching this problem in terms of Carnap's system: (1) Our language can consist of a single family of six predicates, one for each of the possible outcomes of the roll of a single die, and individual constants r_i and g_i denoting the ith roll of the red die and the ith roll of the green die respectively; (2) the language may consist of two individual constants r and g denoting the red and green die respectively, and several families of six predicates F_i^j such that "$F_i^j(r)$" means the same as "r has the property of having resulted in i on its jth toss; (3) the language may consist of the thirty-six "compound"

[12] *The Philosophy of Rudolph Carnap*, ed. by P. A. Schilpp (LaSalle, Illinois, Open Court, 1963), pp. 966–979.

predicates which describe the possible results of a roll of the pair of dice and individual constants a_i which name the outcome of the ith roll of the pair of dice. On this approach "$F_i F_j(a_k)$" would mean that on the kth roll the red die has i on its uppermost face and the green die has j on its uppermost face.

The numerical results obtained for the above examples can be *exactly* duplicated only for the third approach, although basically the same points can be made using the other two approaches on the basis of slightly different numerical values. It can be shown, however, that the third approach is the only one which leads to *prima facie* intuitively acceptable values. Hence we will adopt it for the purpose of exposition.

Axioms A_7 and A_8 of Carnap's criteria of adequacy for definitions of what he calls "logical or inductive probability" state that $P(p) = P(q)$ if p can be obtained from q by a permutation of the non-logical constants of the language, where 'P' stands for the measure of logical probabilities in the absence of any (relevant) observational data (it is the explicandum of his "m-functions").[13] Hence for the third approach to the dice example, $P(F_i F_j(a_k)) = P(F_l F_m(a_n))$ for all possible values of i, j, k, l, m, n. For any fixed k there are 36 incompatible and exhaustive sentences of the form "$F_i F_j(a_k)$" in the language. It follows that $P(F_i F_j(a_k)) = 1/36$ for all possible values of i, j, and k.

Let us suppose that we do not have any observational data concerning the rolls of the dice.[14] Let 'h' stand for the sentence which asserts that sevens will occur on roll i, i.e., the sentence $F_1 F_6(a_i)$, or $F_2 F_5(a_i)$, or $F_3 F_4(a_i)$, or ... $F_6 F_1(a_i)$," which has a P-value of 6/36 since the six disjuncts are incompatible. Let 'e' stand for the sentence which asserts that at least one 3 has been obtained on roll i. Now e is the disjunction of 11 incompatible sentences $F_j F_k(a_i)$ such that j or $k = 3$, and hence has a P-value of 11/36. The conjunction of h and e is the sentence "$F_3 F_4(a_i)$, or $F_4 F_3(a_i)$," with a P-value of 2/36. Thus as before $P(h/e) = 2/11$. In fact all the calculations for the dice example can be duplicated. So can the arguments which show that the rule of inductive behavior characterized in the opening paragraph leads to inconsistencies, and which suggest that $P(h/e)$ is not a correct guide to inductive behavior concerning h.

---

[13] *Ibid.*, p. 975. This corresponds to the assumption of symmetry in Sect. 3.

[14] Alternatively, we could suppose that the 36 possible outcomes of a single roll have occurred with equal frequency for the rolls that we have observed. Then 'P' would represent the measure of logical probability given the observed results: the values of P would be unchanged according to A_7 and A_8 and the definition of conditional probability.

Carnap's system fares no better for the card example. Let the individuals a_i be the ith hand dealt, and the predicates of the system be the 1,680 predicates which correspond to the 1,680 possible hands of 4 cards. Thus for each individual a_i there are 1,680 incompatible and jointly exhaustive atomic sentences which are equiprobable according to A_7 of Carnap's axioms; hence their P-value is 1/1680.

Let h stand for the sentence which states that Player X has two aces; e for the sentence which states that Player X has an A of H, and Player Y has a K of H and a K of D; and e' for the sentence which asserts that Player X has an A, suit unknown, and Player Y has a K of H and a K of D. $P(e) = 20/1680$ since e is equivalent to a disjunction of 20 of the 1680 atomic sentences involving some arbitrary hand i. $P(h \ \& \ e) = 12/1680$ hence $P(h/e) = (12/1680)/(20/1680) = 12/20 = 3/5$. e' is a disjunction of 56 atomic sentences involving a_i and the conjunction of h and e' is a disjunction of 24 such sentences. Hence $P(h/e') = (24/1680)/(56/1680) = 3/7$. Thus we have exactly the same values as before.

The dice and card examples undermine Carnap's thesis that his system represents an explication of inductive probability. For, as he has clearly seen, the inductive probability of an hypothesis h for a decision-maker whose total relevant observational data is expressed by e—the degree of belief in h which it is rational for him to entertain—is that probability involving h and e which is a guide to his inductive behavior concerning h.[15] Carnap's entire system is based on this insight. However, in formulating his system he proceeded under the assumption that Rule C is sound: he used this rule to determine the inductive probability of h for situations of the type under consideration by, in effect, setting down a recipe for measuring the unconditional probabilities of e and the conjunction of h and e, and then identifying the inductive probability of h with the probability of h given that e is true, i.e., with $P(h \ \& \ e)/P(e)$.

7. Inductive Probability and Ascertaining: Why Rule C is Wrong

In this section it will be shown by means of the dice and card examples that the probability which should be a guide to inductive behavior for situations of the sort described in the first paragraph of this paper is the probability of h given that the data expressed by e were obtained by the decision-maker by such-and-such a manner of collecting, categorizing, and (perhaps) reporting. This is the prob-

[15] Schilpp, *op. cit.*, p. 967.

ability of *h* given that the truth of *e* has been ascertained by the decision-maker.

It is obvious that whenever observational data are obtained they are obtained by some manner or procedure of investigation. It is equally obvious that in order to compute the probability of an hypothesis *h* given that the data expressed by an observation report *e* were obtained, one must know or assume something about the processes and procedures by which these data were obtained. Let us consider some of the methods by which the dice roller could have decided which information to offer us for betting purposes.

Method 1. After the dice are rolled the roller selects one of the dice in a "random" way such that the probability of selecting the red die on any trial is p. He then reports to us the number on the uppermost face of this die without telling us which die it is. He decides that whatever he rolls, his report will be of the form: "At least one x was rolled" (he can always find a true report of this form).

Suppose that the roller reports that at least one 3 was rolled. Let *e* stand for the sentence "At least one 3 was rolled." The probability of our having ascertained *e* will now be: $p \cdot P(3$ on red die$)+(1-p)\cdot P(3$ on green die$)=p\cdot(1/6)+(1-p)\cdot(1/6)=1/6$. Let *h* name the hypothesis that sevens was rolled. The probability of *h* given that *e* was ascertained $= P(h$ & that $-e$ was ascertained$)/P($that $-e$ was ascertained$) = [p \cdot P(3$ on red & 4 on green$)+(1-p)\cdot P(4$ on red & 3 on green$)]/(1/6) = [(1/36)\cdot(p+1-p)]/(1/6) = 1/6$.

The measure of the probability of *h* given that *e* was ascertained clearly agrees with our earlier intuitive assessment of the fair betting odds against *h*. In fact, it is reasonable to suppose that in making that assessment we unconsciously assumed that the roller selected the information he offered us in such a "random" way.

It is worth noting that, making suitable changes in Method 1, the probability of sevens given that it was ascertained that the red die has 3 on its uppermost face $=(p\cdot(1/36))/(p\cdot(1/6))=1/6$, which agrees with what our discussion of the dice example suggested that the inductive probability should be for such an observation report.

The probability of sixes given that it is ascertained that at least one x was rolled will be:

$$\frac{p\cdot P(x \text{ on red } \& \text{ } 6-x \text{ on green})+(1-p)\cdot P(6-x \text{ on red } \& \text{ } x \text{ on green})}{p\cdot P(x \text{ on red})+(1-p)\cdot P(x \text{ on green})}$$

But this will be 1/6 for $x=1, 2, 3, 4$, and 5. Again, this result is in perfect agreement with the argument which suggested that the inductive probability of sixes for a decision-maker who is informed by the roller that at least one x was rolled is the same (namely, 1/6) for $x = 1, 2, 3, 4$, and 5, in contradistinction to what Rule C recommends.

Of course all the "correct" values obtained above depend on the assumption that Method 1 was used by the roller. Quite different probability values may be obtained on the basis of other methods. Consider the following:

Method 2. The form of the report is the same as for Method 1, but the roller selects as the value of x the largest of the two numbers rolled.

Let e_x state that at least one x was rolled. It can be seen that the probability of sevens given that e_x has been ascertained is 0 for $x = 1, 2$ & 3; 2/7 for $x = 4$; 2/9 for $x = 5$; 2/11 for $x = 6$.

Note that Rule C doesn't even hold for "direct," "non-disjunctive" observation reports: Suppose that the roller decides to present a report of the form "x is on the - - - die," and the die he will choose to report about is that die which has the largest of the two numbers rolled, and the red die in case of a tie. Letting r_x denote "x is on the red die" and g_x denote "x is on the green die," it can be seen that (a) $P(\text{sevens}/r_x$ has been ascertained) $= P(\text{sevens}/g_x$ has been ascertained) $= 0$ for $x = 1, 2$, and 3; (b) $P(\text{sevens}/r_x$ has been ascertained) $= 1/4$ for $x = 4$, 1/5 for $x = 5$, and 1/6 for $x = 6$; (c) $P(\text{sevens}/g_x$ has been ascertained) $= 1/3$ for $x = 4$, 1/4 for $x = 5$, and 1/5 for $x = 6$. Thus in order to be entitled to assign an inductive probability of 1/6 to sevens if the roller tells us that, say, a three was rolled on the red die, we have to be justified in assuming that this information was not arrived at by a procedure like the one above.

Method 3. We ask the roller if at least one x was rolled, in order of increasing values of x. He (truthfully) answers "Yes!" or "No!" We stop asking when a "Yes!" is obtained. Under Method 3 the probability of sevens given that e_x has been ascertained $= 2/11$ for $x = 1$; 2/9 for $x = 2$; 2/7 for $x = 3$; and 0 for $x = 4, 5, 6$.

Method 4. We ask *once* a question of the form "Is x true?" for some definite value of x we have decided upon. The roller will, we assume, answer truthfully. Under Method 4 the probability of sevens given that e_x has been ascertained is 2/11. The probability of sevens given that we receive a "No!" answer is 4/25. Note that under this method the "incorrect" probability value obtained in Sect. 3 is in fact equal to the probability which should be a guide to inductive behavior. (I must confess that this puzzles me a bit.)

Are we justified in evaluating fair betting odds for the hypothesis that sevens was rolled on the basis of a report from the roller, in the absence of any knowledge of and/or assumptions about the procedure used by the roller in "generating" his report? Are we justified in assessing the inductive probability of this hypothesis in a state of "complete ignorance" about the procedure? The arguments of Sect. 3 and the results just obtained seem to clearly show that the answer is "No!"—regardless of whether or not the report is "direct" or "indirect," "conjunctive" or "disjunctive" (with the exception of cases in which sevens cannot occur if the report is true); they seem to show that to plead ignorance about the procedure is tantamount to declaring that one is not in a position to assess the evidential force of the report.

We are now in a position to see why Rule C is incorrect: The inductive probability or degree of credibility of an hypothesis for a person at a given time does not depend *solely* (as Rule C presupposes) on its semantic or logical content relative to that of a sentence which he takes as an accurate expression of all the relevant observational data at his disposal (and on the background information, if any, which provides the basis for a probability distribution over the possible observations); it also depends on the procedure and underlying processes—always present in a practical situation, known or unknown—by which the data were generated. A discussion of the card example may make this point a bit clearer.

Let o state that I have a K of H and a K of D. Let e state that you have an A of S, and let e' state that you have an A. Let h stand for the hypothesis that you have a pair of aces. The probability of h given that I have ascertained the truth of o and of $e' = P(h/o$ and your first card is A and my spy sees it, or o and your second card is an A and my spy sees it). Let f abbreviate "My spy sees the first card dealt to you," and f' "My spy sees the second card dealt to you." Let p be the measure of the probability of f. We assume that either f or f'. Hence the probability of h given that I have ascertained the truth of $e' =$

$$P(h \,\&(f \text{ or } f')/o)/P((A \text{ on 1st } \& f) \text{ or } (A \text{ on 2nd } \& f')/o) = (12/30)/((p \cdot 20/30)+(20/30)(1-p)) = 3/5 \ .$$

The probability of h given that e and o have been ascertained by me will be:

$$\frac{P(h \,\& ((A \text{ of } S \text{ on 1st card } \& f) \text{ or } (A \text{ of } S \text{ on 2nd } \& f')/o)}{P((A \text{ of } S \text{ on 1st } \& f) \text{ or } (A \text{ of } S \text{ on 2nd } \& f')/o)} =$$

$$\frac{p \cdot (3/30)+(1-p) \cdot (3/30)}{p \cdot (5/30)+(1-p) \cdot (5/30)} = 3/5$$

These results agree with the conclusions reached in the earlier discussion of this example.

The values just obtained depend on the assumption that the order with which the card my spy can see is dealt is stochastically independent of my opponent's hand and also that it is independent of the cards in my hand. The former assumption may be false due to certain peculiar habits my opponent has in arranging his hand. Or both of these assumptions may be false because my opponent is quite sure that I have a spy with a high powered telescope and he has a spy of his own. The values just obtained obviously would not hold if such were the case. The main point is that I can compute inductive probabilities for this case only if I am willing to make *assumptions* about my opponent's habits, honesty, and degree of suspiciousness. Furthermore, the correctness of my assessment of the degree of credibility of h in the light of the spy's report depends on the *truth* of the assumptions I adopt *in order* to make this assessment. If the assumptions are not true, the assessment is not *objectively* correct, although if they are reasonable for me to make, the assessment is *subjectively* correct. The same holds for the dice example and for most, if not all, real-life decision-making problems of the sort described in the opening paragraph of the paper.

The reader will note that not one point made in connection with the examples has anything to do with questions concerning the *reliability* of the processes and procedures by which our observation reports are generated; it was simply assumed in the discussion of the examples that they are reliable, i.e., that the reports involved are true. Of course the amount of information contained in a given report is a function of the reliability of the procedures by which it was generated; but it also depends upon what the procedures actually *are*, regardless of their degree of reliability, and this is what I am trying to show.

8. Probability and Conditional Bets

Some readers may find it difficult to accept the arguments presented above because of a conviction that all probabilities must have something to do with the guidance of "betting" behavior, by virtue of the meaning of probability. They may feel that the arguments must be wrong because it is inconceivable that, say, the probability of sevens given the truth of "At least one three has been rolled" is totally worthless as a guide to betting behavior, as opposed to the probability of sevens given that it has been ascertained that at least one three was rolled. But this has not been asserted. It seems quite clear that any probability worthy of the name must have something to do with the guidance of some form of betting behavior. Consider the dice example. The probability of sevens given that at least one three has been rolled is a guide for bets *conditional* on at least one three being rolled, but not for bets in which the bettor has ascertained that at least one three has been rolled. A bet on h conditional on the occurrence of e is a bet such that if e does not occur no money is exchanged, but if it does occur, the monies bet are exchanged. Thus for a bet on sevens (h) conditional on the occurrence of at least one three (e) if X bets \$$s$ on h, and Y bets \$$t$ on not-h, then if e and h occur Y gives X \$$t$ and if e and not-h occur X gives Y \$$s$, and if the conditioning event does not occur the bet is off. Fair odds against h for such a bet can be obtained by finding that value of t/s for which the expected value of a bet on h (or not-h) is equal to zero. The expected value of a bet on h is $(P(h \ \& \ e) \cdot t) - (s \cdot P(e \ \& \ \text{not-}h)) + P(\text{not-}e) \cdot 0$. Setting this equal to zero we get $t/s = P(e \ \& \ \text{not-}h)/P(e \ \& \ h) = P(e) \cdot P(\text{not-}h/e)$ divided by $P(e) \cdot P(h/e) = (1 - P(h/e))/P(h/e)$. For the dice example $P(h/e) = 2/11$, hence fair odds against h for a bet conditional on at least one three being rolled are 9 to 2.[16]

One can readily see from the above that in general the probability of h given the truth of an observational report e is in fact a guide for bets on h conditional on e's being true. It is clear that such bets do not always correspond to decision-making problems of the sort—described in the first paragraph of the paper—that inductive logic is concerned with. It is worth noting that Carnap cites Kemeny's proof[17] that a person's betting behavior is "coherent" if and only if his subjective probabilities (as exemplified by the odds he is willing to accept for bets) conform to the usual principles of probability as his justification for defining the measure of the logical probability of h relative to an observation report e by the usual formula for

---

[16] Some may consider odds of 9 to 2 paradoxical even in the context of conditional bets. It may be felt that when the dice are rolled some instance of "At least one x was rolled" is true, and since $P(h/\text{at least one } x \text{ was rolled}) = 2/11$ for all values of x, the probability which should be a guide to betting behavior for bets on h which are conditional only on the dice being rolled (which is 1/6) should be the same as for bets conditional on some given instance of "At least one x was rolled." This feeling would be justified if the instances of "At least one x was rolled" are logically incompatible as is the case for "x was rolled on the red die." For if $p_1, p_2, \ldots, p_n$ are incompatible, equiprobable and such that $P(q/p_i)$ is the same for all i, then $P(q/p_1 \text{ or } p_2 \text{ or } \ldots p_n) = P(q/p_i)$. However if one reflects on the compatibility of the different instances of "At least one x was rolled," the air of paradox should vanish.

[17] "Fair Bets and Inductive Probabilities," *The Journal of Symbolic Logic*, vol. 20 (1955), pp. 263–273.

conditional probability. But in his proof Kemeny shows only that this formula holds for conditional bets. Both he and Carnap have failed to see that the probabilities which are appropriate for such betting situations are not always appropriate for the "betting" situations involved in the typical case of inductive behavior. This is not to claim that conditional bets *never* correspond to such cases. For, the probability which is a guide to inductive behavior for such cases—the inductive probability of the hypothesis for the decision-maker—is that probability which determines fair odds for bets which are conditional on the decision-maker *ascertaining* the truth of the observation report. All conditional probabilities are fair betting quotients for conditional bets, but only those conditional probabilities which involve the ascertainment of observational data can serve as guides for conditional bets that correspond to situations of the sort described in the first paragraph of the paper.

9. "Total Evidence" and Observational Reports

One is not justified in concluding that Carnap's *approach* to inductive probability is inadequate *simply* on the ground that the inductively inappropriate probability values associated with the dice and card examples are deducible from his axioms for inductive probability. Nor is one justified in simply rejecting Rule C on the basis of the arguments of Sects. 3 and 4. For it can be plausibly argued that (1) as Carnap has emphatically pointed out, if a person ". . . wishes to apply a principle or theorem of inductive logic to his knowledge situation then he must use as evidence *e* his total observational knowledge"[18] and (2) in the calculations of Sects. 3 and 4 this requirement of total evidence was violated, whereas the "correct" values were obtained in Sect. 7 because this requirement was applied.

But what relevant information, "observational" or otherwise, available to the bettors, the evidence-weighers, was omitted in the calculations of Sects. 3 and 4? The only possible candidates are (a) in the case of the dice example, that the roller *volunteered* the information that at least one three was rolled, and (b) for the card example, that I was told by my spy that he could see only part of one card, just enough to see that it is an ace but not its suit. Now this sort of information is customarily regarded as relevant by proponents of Rule C only when there is some question of the reliability of a report, which is not here at issue. But disregarding this, how can such

[18] *The Philosophy of Rudolph Carnap, op. cit.*, p. 972.

information be processed in the calculation of fair betting odds? The answer is clear. For the dice example, *only by assuming* (guessing) something about the form, if any, the roller had decided upon for the report to be transmitted and something about the procedure he employed for filling this form out.[19] And for the card example, *only by assuming* something about the *underlying mechanism* or *process* involved in the "selection" of the card my spy gets to see. Thus to hold that relevant *information* was omitted in the first set of calculations for the dice and card examples but was brought into the calculations of Sect. 7 is to grossly misunderstand the structure of the examples.

Another possible line of defense of Rule C and Carnap's axioms is to claim that the axioms are not *designed* to handle cases where part of the available observational data includes information about who reported the observations and/or some of the circumstances under which they were made, and that Rule C failed in Sects. 3 and 4 for the same reason, namely, that we simply didn't have a probability distribution over the propositions which really represent all the possible total observation reports, so that the dice and card examples only show that Carnap's *present* systems are more incomplete than has been realized and not that they are wrong in principle. This line of defense won't work because (a) in practice we almost never obtain our observational data by simply asking an articulate and benevolent Mother Nature "Is such-and-such the case?" and getting a "Yes" or "No" answer in the manner of Method 4, and (b) we almost always need to make *particular assumptions*, applying *specifically* to the problem-context at hand in order to *justify* the assignment of the requisite probability distribution; hence there seems to be no way to extend Carnap's present systems to deal with these cases. There is little difference between being wrong in principle and hardly ever applicable.

Rule C and the Carnapian approach to inductive probability are basically speculative applications of a widely held (among philosophers) modern version of the traditional empiricist thesis that all our empirical knowledge is derived from sense-impressions and only sense-impressions. According to this version of empiricism the cognitive worth (degree of credibility, degree of confirmation) of an empirical hypothesis at a given time is a determinable function of solely its *semantic* or *logical* content relative to that of a sentence

---

[19] Or by assuming that the form and procedure used are one of a given list, and assigning a probability distribution over these which represents the likelihood of employment. Similar remarks hold for the card example.

which expresses all the available relevant "observational" data.[20]

Despite its apparent obviousness this view is, or at least seems to be, incompatible with the cumulative wisdom of practicing statisticians and scientific researchers. Nevertheless, proponents of it have either explicitly rejected the counsel of this wisdom[21] or have casually taken the position that the incompatibility is merely apparent, that the presence of vast amounts of background knowledge in scientific contexts obscures the logical structure of the weighing of evidence in these contexts, but that since this structure is essentially the same in very simple cases—where the view holds—the practical wisdom of the experts (although not perhaps their theory) can be incorporated within its framework. This paper represents an attempt to show that the reasons for the apparent failure of this view in scientific practice arise even in the simple and well-defined kinds of cases its proponents cite as paradigms, and hence that it is wrong because it embodies an incorrect view of the structure of the weighing of evidence.

An adequate account of the weighing of evidence must acknowledge the following facts: The evidential weight of observational data essentially depends on the manner or process, known or unknown, deliberately contrived or natural, by which or under which data are collected, categorized, and reported. Evidence-weighers usually do not possess the required information about these factors—factors which must be brought into calculations of the evidential weight of data if this is to be properly evaluated *at all*. (I am not here talking about the frequent need in practice to assess the accuracy of data.) Thus even if a report e—"observational" or otherwise—expresses all the available "knowledge," "data," or "information" (call it what you will!) which is relevant to an hypothesis h, it is often the case that the degree to which e supports h—the amount of information contained in e with respect to h—cannot be determined and can only be estimated on the basis of assumptions about the process by which

[20] Some insist that the data be "certain," others not. The arguments raised against this view in the paper apply equally to either position. It is worth noting that this view not only underlies Carnap's approach to inductive probability, it also underlies the confirmation theories of Hempel, Oppenheim, and Kemeny, to mention just a few.

[21] For example, on p. 494 of *Logical Foundations of Probability*, 2nd ed. (Chicago, University of Chicago Press, 1962), Carnap rejects the time-honored view that the hypergeometric distribution applies to a sample only if it is random, on the ground that (1) "... we can hardly ever know whether a sample is a random sample in the defined sense," and (2) "The validity of any inductive inference, even in practical application, does not depend on the actual state of affairs, and certainly not on any unknown frequencies; it depends merely on the given knowledge situation or, more exactly speaking, on the logical [i.e., semantic —my addition] relations between the given evidence and the hypothesis."

the observations expressed by *e* were generated. In addition, many descriptions of the states of affairs about which assumptions have to be made in order to weigh evidence can be labelled "observational" only by stretching the range of application of this term so far beyond its accepted boundaries as to render its usage empty of significance.[22] But while such descriptions are not "observational" they can be empirically *tested* should the need to justify them arise . . . on the basis of other assumptions.

Lehman College of the City University of New York

---

[22] Consider the dice and card examples. One could hardly characterize as observational sentences to the effect that (a) the dice-roller's selection of a die is *independent* of the numerals rolled (Method 1), or (b) that he has decided that his report will be of such-and-such a form (Methods 1 & 2), or (c) that the question I ask is *independent* of the results of the roll (Method 4), or (d) that my spy's seeing the first card dealt is *independent* of my opponent's hand.

VIII
Popper on Learning From Experience

JOSEPH AGASSI

SIR Karl Popper has claimed repeatedly that different events are independent of each other in the sense of "independence" used in the theory of probability. That is to say, the probability of a conjunction of two different events is the product of their probabilities. From this it follows that the probability of any event given any set of different events is the same as it was prior to those events having been given. For example, the chance of the next swan being white is not affected by our having seen many swans before, all of which are white. Similarly, Sir Karl has insisted that the initial probability of any universal law is zero. From this it follows that the probability of any universal law is always zero, regardless of how much empirical evidence supports or backs it.[1]

This should lead to a complete breakdown of rational science. If Sir Karl is right, then, it seems, there is no mode of rational choice of scientific theories; and with this the hopes that we do—or at least may—learn from experience must evaporate.

All this worried Diane, and she has asked for a clarification of his view on this point: how does Sir Karl view learning from experience? I have set my clarification in a few preliminary paragraphs which break into an imaginary dialogue between Diane and Sir Karl. Though I do think that the dialogue is faithful in content, and though I have chosen the atmosphere with as much care as I could, I am afraid I have taken a literary license here and there and have made no attempt to make the dialogue faithful in every respect to Sir Karl's characteristic style of conversation. Hence, this work has no claim for great biographical accuracy.[2]

Customarily, philosophers of science discuss at length the problem of choice of scientific theories. The words "choose," or "choice," and the frequently used expression "we choose," are ambiguous. By "we" a philosopher often means Einstein and himself. By "choose"

---

[1] The points raised here are developed further in the Appendix.
[2] Also, the style of the dialogue is less conversational than originally intended, because editors and referees exercise not only philosophical judgments but also artistic and stylistic ones. Lynn Lindholm made some final improvements.

he means, *believe to be true.* Thus, "we choose general relativity" usually means Einstein and the philosopher believe that general relativity is true. This, of course, is utterly false: Einstein disbelieved general relativity on account of some metaphysical arguments, and the philosopher all too often does not believe general relativity since he cannot believe a thesis he does not know and all too often he does not know what general relativity says. So there—refuse to understand "we choose" and maybe you will have nothing more to understand or fail to understand.

To set the problem a bit more cautiously. Men of science are often faced with alternative hypotheses, and they try—sometimes successfully—to make an experiment help them choose one, i.e., take one to be the one which they tell their students and lay-audiences to choose, or at least the one which they talk more about. Also, some say, it is the one hypothesis which they train engineers and navigators and their likes to use. This is ridiculous because engineers and navigators use Newton, not Einstein—excepting some nuclear engineers and some space navigators (and even then, they are more likely to use special relativity, hardly ever general relativity).

Never mind. Suppose we have a set of competing hypotheses. (What Makes Newton and Einstein, Lamarck and Darwin, but not Lamarck and Einstein, competitors? If you find the answer keep it in mind.) Suppose under some conditions experience helps us choose. Which conditions? What rules of choice? The process of choice may be called inductive. The rules may be called inductive logic.

Now, the Carnapian, or current, theory may be characterized simply as follows. Inductive logic follows the rules of the calculus of probabilities. The conditions of choice are two: (1) We must set a measure of *a priori* probabilities and dependences. These are not specified by the calculus, and must be added. As we shall see, this is easier said than done. (2) We may now bring relevant evidence to tip probability in favor of one candidate against the rest.

By contrast, and the accent is on contrast, the Popperian theory is this. In science the important process is not at all choice or endorsement but rather criticism or rejection, namely the conclusion that a given theory is unsatisfactory in view of this, that or another specific criticism—usually specific empirical data which (seem to) conflict with the theory.

Q. Do scientists choose?
A. Usually, yes.
Q. Do they have to?
A. Not really.

Q. Is their actual choice rational?
A. Yes.
Q. How so?
A. They do reject unsatisfactory theories, and so what they do not reject they have failed to declare unsatisfactory, i.e., they have corroborated.
Q. Is this a logic?
A. To some extent, though not all the way.
Q. Can we call this logic inductive?
A. Call anything by any name.
Q. Sorry. I mean, is there a problem of choice and is the problem soluble by empirical means and a logic and have you not supplied the logic?
A. No.
Q. How so?
A. It is not the case that scientists have a problem of choice and we philosophers of science offer them tools. Rather, it is the case that scientists do choose. Their choice may be extrascientific or it may be anti-scientific, but it cannot be scientific—it is always metaphysical, though more or less in accord with science (more or less in accord with my theory of corroboration).
Q. The distinction is subtle: you simply refuse to include the metaphysics as a part of the logic of science. Why?
A. Because the problem of induction is not that of how we choose a hypothesis, but that of how we learn from experience. To the second question "How do we learn from experience?" the current answer is by choice. And it is this very answer which leads to the first question "How do we choose?" If you do not ask the second question, or if you do not give the second question the current answer, you need not ever bother with the first question. And, indeed, the current answer is untrue. We learn from experience not by choice, but by rejection. And this answer leads not to the question "how do we choose?" but to the question, "how do we reject?"
Q. If so, why did you, Sir Karl, study choice?
A. Choice does occur, and it is not inductive, as I have tried again and again to show.
Q. Why not inductive?
A. Inductive logic is the logic of learning from experience by choice and it follows the rules of the calculus of probability. Corroboration is somewhat a logic of choice, but not a logic of learning

from experience and it does not follow the rules of the calculus of probabilities.
Q. So you agree with Carnap: his logic does and yours doesn't follow the calculus of probabilities, and so they differ?
A. Obviously.
Q. Do your views do not compete?
A. Our *logics do differ*; this is why our *views do clash*.
Q. How so?
A. In my view scientists choose in accord with my logic. In Carnap's view scientists choose in accord with his logic. Our logics differ: we both apply them to the same phenomena: therefore, we arrive at a conflict. Q.E.D.
Q. Not so fast. Can you not apply different logics in different ways so as to get the same results?
A. Yes, you can. You are quite right.
Q. In which case there will be no different views?
A. Correct again.
Q. Can this be the case of Popper versus Carnap?
A. No.
Q. Can you prove this?
A. Easily: Carnap's logic as applied by Carnap, when applied to choices in scientific situations leads to some wrong results: Popper's only to right results. Hence they differ. (Even if they are both ultimately mistaken, they still differ because the one is now known to be mistaken but not the other.)
Q. Please spell this out.
A. You are tiresome: all you need to do is read my works, they are crystal clear.
Q. I deny that vehemently. Don't go away! Please explain.
A. As Carnap says in his *Continuum of Inductive Methods*, if the next event depends in no way on the last few observed ones, our last observations are no guide to the choice of a guess on its outcome (by definition of independence). Even the observation of a million white swans does not tell us, in this case, what is the color of the next swan, and *a fortiori*, what is the color of all (unobserved) swans. And therefore, as Carnap notes, probabilistic independence prevents experience from helping us choose. Hence we do not choose by Carnap's rules unless there is dependence. Carnap says we do choose by my rules, hence that there is dependence. Let it be so. What measure of dependence? Should we assume that measure to be high or low? That is to say, at what speed do we learn from past experience? Carnap says

that he doesn't know. He wants to consult experience about this. This is funny: he wants to consult experience about the rules for consulting experience, which is either a vicious circle or infinite regress. Also, we may learn fast in one field and slowly in another field. How will Carnap find this out? By experience? But experience may be balanced by unequal distributions of intellectual energy. Carnap correlates intuitively worlds with higher measures of dependence in them, namely worlds with more order in them, with ideal students learning more rapidly from experience, namely quicker to generalize, and less ordered worlds with ideal slow learners. But learning is supposed to tell us chiefly how and to what measure the world is ordered! Moreover, probabilistic dependence leads under the best conditions merely to a choice of *limited forecasts*, but in a *possibly infinite world* (in space or time) *even dependence does not lead to choice or universal statements:* even a big measure of dependence fails to let experience rescue universal statements from their initial zero probability.

Q. But if there is dependence not all universal statements have zero probability.

A. You mean I have made a small logical error?

Q. Did you not?

A. No. There can be dependence in an infinite sequence of event-statements the conjunction of all of which is equivalent to one universal statement. And there can be dependence in an infinite sequence of universal statements; these two possibilities are mutually (logically) independent.

Q. Good Lord!

A. That is right. You try to solve a simple question of choice by a simple calculus of probability, and the calculus rebels and turns up in a really complicated fashion.

Q. Why? Is it as complicated in mathematics?

A. Not in such an annoying way.

Q. Meaning?

A. Questions of dependence and of probability measures are decided in mathematics arbitrarily, just as in the case of arbitrarily different geometries, and their consequences can be studied.

Q. And in statistics, how do we determine which of these arbitrary probability measures to use?

A. Much as we decide which arbitrary geometry to apply.

Q. Namely?

POPPER ON LEARNING FROM EXPERIENCE

- A. We try any of them for size and if it fits—
- Q. —we choose them?
- A. —we test them—
- Q. —and then choose them?
- A. —and try to eliminate them—
- Q. —and if the tests fail we choose them?
- A. —and if the tests fail we *may* choose them.
- Q. —Oh!
- A. The probability of the inductive philosophers is that of compulsory choice. But there is no compulsory choice. And so philosophers need no probability, and thus no probability measure, and no dependence measure. Rather, for different problems in probability theory we assume alternating solutions involving different measures of probability and of dependence.
- Q. So we need not say that all events are *a priori* independent and all hypotheses have zero initial probability!
- A. You are right.
- Q. So why do you say what we *need* not say?
- A. I say what *may* be said.
- Q. On what ground?
- A. On the ground that experience and the laws of probability make me choose it.
- Q. You must be pulling my leg.
- A. Indeed, I am.
- Q. Why do you tease me? Do you have no heart?
- A. I do. It is because I do, and because you try to make me apply induction to my choice of equiprobability of events and to my choice of zero initial probability of hypotheses, that I say what you wish me to say.
- Q. Sorry. What criteria, other than inductive, do you employ for the choice of independence and zero probability?
- A. I do not fully know, but perhaps simplicity is one.
- Q. But how can you call your choice of zero probability and total independence simple if it makes life so difficult; for does it not prevent any further choice, in particular choice of empirical hypotheses or learning from experience?
- A. Take care! Now you are committing a small logical error. My choice is legitimate and excludes inductive logic. But it permits, to begin with, learning from experience by rejection and after this, as an option, the choice of hypotheses by corroboration.
- Q. And is this inductive or not? I am at a loss.
- A. As you yourself say, my dear, zero initial probability of hy-

potheses plus equiprobability of events prevents inductive logic, but not the logic of rejection and of choice by corroboration.

Q. Can you prove the latter point—on corroboration, I mean?

A. Certainly; with mathematical precision, even.

Q. And is it not less simple to say that learning from experience is by rejection but that choice is corroboration?

A. Less simple than what?

Q. Than saying that *both* learning from experience *and* choice are corroborations?

A. This is an inductive variant of Popperism.

Q. I am delighted. What's wrong with it?

A. It does not work.

Q. Oh! Why?

A. Infinite regress and all that. Good night.

Q. Wait! You didn't say yet. Why do you assume events to be independent?

A. I do not. They interdepend in accord with causal laws.

Q. I mean why do you assume events to be *a priori* independent?

A. I do not. I assume *a priori* that some depend on others some not, in accord with. . . .

Q. Sorry again; why is the probability of events in your system such that they are probabilistically independent of each other? Is that precise enough for your finicky taste?

A. Almost. You are very acute, if I may compliment you. I say, events are probabilistically independent in the absence of laws.

Q. Thanks. Why?

A. Because laws and only laws are the measures of the mutual dependence of events.

Q. How interesting! And for Carnap both laws of nature and *a priori* probabilities are measures of dependence of events. Is that why you say your system is simpler than Carnap's?

A. This is, roughly, the general idea.

Q. Any third alternative?

A. I may not be overjoyed to agree with Carnap, but I am afraid I must. No third alternative is logically possible. Only two inductive systems are possible. Either you assume *a priori* dependence in your system, or you do not. The former is his, the latter is Wittgenstein's and mine. You see, I even have to agree with Wittgenstein.

Q. And Carnap's system is defective, whereas in yours no learning from experience is at all possible!

A. Not so fast or you have again a small logical error on your hands.

What you should have said is this. In Carnap's system no kind of learning from experience is possible, and in mine no inductive learning from experience is possible. Good night.

Q. Please wait. But learning from experience by elimination is possible?
A. By elimination of errors, not by confirmation of one view through the elimination of another. Good night.
Q. Please, wait. One last question, please. Just how do we learn from experience by refutation?
A. Good night.
Q. Are you peeved?
A. Yes.
Q. Why?
A. All the time I told you again and again that we learn from experience by refutations and you didn't bother to try to understand.
Q. But I was bothered with another point. Can I bother about all points at once?
A. No. But why bother with inductive logic for so long before even trying to see what I mean?
Q. What *do* you mean by learning from experience?
A. I mean our intellectual horizon widens with a refutation. I mean that a refuting observation report is more theoretically loaded the more abstract and general the theory it refutes. I mean that when a theory is refuted we may see better how far it goes, explain it, and so see perhaps why it goes that far: some breakthroughs of great importance are great refutations.
Q. Is this a theory of breakthrough?
A. No. Breakthroughs are unique; get them anyway you like. If you can't, try refuting an existing theory.
Q. Which one?
A. Good night.
Q. Peeved again?
A. No. Tired. Inductivists think that all repetition is reinforcement. I wonder. Sometimes repetition simply fills one with profound tiredness, if I may make an empirical observation. But I know: philosophers are not supposed to take recourse to experience. Good night.

Boston University

Appendix

Definition of independence: a and b are independent when and only when $p(a \& b) = p(a) \times p(b)$.

Definition of conditional probability: If $p(b) \neq 0$, then $p(a, b) = p(a \& b)/p(b)$.

If a and b are independent, then
$p(a, b) = p(a \& b)/p(b) = [p(a) \times p(b)]/p(b) = p(a)$.

If $p(a) = 0$ then, by the law of monotony, $p(a \& b) = 0$, and $p(a, b) = 0$.

If $p(b) = 0$, then $p(a, b)$ is undefined for most systems (including Carnap's). It is normally assumed in the literature that since b is an observation or observation-report of unique events, $p(b)$ is never zero. This seems to me to be very questionable. Assume $p(b)$ to be zero, and the computing of $p(a, b)$ may be highly problematic.

Bibliography

Joseph Agassi, "The Role of Corroboration in Popper's Methodology," *Australasian Journal of Philosophy*, vol. 39 (1961), pp. 82–91.

―― "Discussion: Analogies as Generalizations," *Philosophy of Science*, vol. 31 (1964), pp. 351–356.

Y. Bar-Hillel, "Comments on 'Degree of Confirmation' by Professor K. R. Popper," *The British Journal for the Philosophy of Science*, vol. 6 (1955–56), pp. 155–157.

Rudolph Carnap, *The Continuum of Inductive Methods* (Chicago, University of Chicago Press, 1952).

―― *Logical Foundations of Probability*, 2nd ed. (Chicago, University of Chicago Press, 1962).

―― *The Philosophy of Rudolph Carnap*, ed. by P. A. Schilpp (LaSalle, Illinois, Open Court, 1963).

Sir Karl Popper, "'Content' and 'Degree of Confirmation': A Reply to Dr. Bar-Hillel," *The British Journal for the Philosophy of Science*, vol. 6 (1955–56), pp. 157–163.

―― *The Logic of Scientific Discovery* (London, Hutchinson, 1959).

―― "On Carnap's version of Laplace's Rule of Succession," *Mind*, vol. 71 (1962), pp. 69–73.

―― *Conjectures and Refutations* (London, Routledge & Kegan Paul, 1963).

IX
Physics and Furniture
D. H. MELLOR

THE proper effect on everyday beliefs of accepting physical theory is as hotly debated as ever, with the most extreme views the most fashionable. On the one hand, there is the Oxford view of Ryle [22] and Strawson [27] that the proper effect is zero; on the other, the American "super-realist" view of Feyerabend (e.g., [4] and [5]), Sellars [24], and Maxwell [11] that the proper effect is one of total replacement.[1] In this paper I relate and reject both views and advocate a modest realist view of physical theory. The problem is not only interesting in itself, but also bears closely on such topics as the relation between philosophical and scientific reconstructions of languages, natural and artificial languages, and the autonomy of analytic philosophy (see e.g., Strawson [27], and Carnap [2]). A satisfactory solution seems to me essential even to formulate these further central and under-analyzed problems correctly.

The problem may usefully be approached via an analysis of Eddington's notorious concept of two worlds, the familiar and the scientific:

> There are duplicates of every object about me—two tables, two chairs, two pens.... One of them has been familiar to me from earliest years. It is a commonplace object of that environment which I call the world. How shall I describe it? It has extension; it is comparatively permanent; it is coloured; above all it is *substantial*.... It is a *thing*.... Table No. 2 is my scientific table. It is a more recent acquaintance and I do not feel so familiar with it. It does not belong to the world previously mentioned.... There is nothing *substantial* about my second table. It is nearly all empty space ... even in the minute part which is not empty we must not transfer the old notion of substance.... I need not tell you that modern physics has by delicate test and remorseless logic assured me that my second scientific table is the only one which is really there.[2]

The scientific world has changed in many details since 1929, but its relation to the familiar world remains for many people the same both

---
[1] Numbers in brackets indicate entries in the list of References at the end of the article.
[2] Eddington [3], pp. xi–xiv.

in interest and unclarity. Much of the unclarity has been dispelled by Stebbing [25], but her comments on Eddington's problem are rather too therapeutic. She is quite right to observe that there are not "duplicates of every object" in the familiar world:

> Eddington firmly excludes *colour* from the scientific world, and rightly so. But the *rose* is coloured, the *table* is coloured, the *curtains* are coloured. How, then, can that which is not coloured duplicate the rose, the curtains, the table?... a coloured object could be *duplicated* only by something with regard to which it would not be meaningless to say that it was coloured.[3] ... It is as absurd to say that there is a scientific table as to say that there is a familiar electron or a familiar quantum.[4]

Thus Stebbing resolves Eddington's dilemma by separating the scientific world from the familiar world. An entity in the one may have a "counterpart"[5] in the other, as an electromagnetic wavelength is the counterpart of a colour,[6] but not a duplicate.

But even when Eddington's confusions have been dealt with, his problem remains. The relation "counterpart of" is not much clearer than the relation "duplicate of." We still need to know how

> the symbolic construction at which physics aims is related to the familiar world. There would seem to be three alternatives: (1) the construction is an imitation of the world; (2) the construction is more real than (or truer than?) the familiar world; (3) the construction is for the sake of correlating certain selected elements in the familiar world, in order that the range of our experience may be extended and what is sensibly experienced may be ordered.[7]

Stebbing's description of the alternatives shows clearly enough that she is going to adopt the third, instrumentalist, view of physical theory. Considered merely as instruments for predicting and ordering experience, perhaps theories raise no problems of truth (since they make no statements of fact), and hence no problems of ontology (since they postulate and describe no entities). But this solution is not available to one who would take even the most modest realist view of theories, for whom there still seem to be two worlds. Even if such a realist agrees that Stebbing has shown these worlds to have no entities in common, that hardly makes the problem of accounting for their relation less urgent.

[3] Stebbing [25], p. 60.
[4] *Ibid.*, p. 58.
[5] *Ibid.*, p. 60.
[6] The example is Stebbing's, *ibid.* Although a common example, it is a bad one, since only for monochromatic light does wavelength correlate with color, and very little light is monochromatic. Other light of the same color may contain *no* light of the corresponding monochromatic wavelength.
[7] *Ibid.*, p. 66.

The first step towards a realist solution is to clarify the sense of the term "world" as it is used in stating the problem. There is the sense in which the world of industry is a different world from the academic world, or an isolated geographical region with exotic forms of life is a different world.[8] The sense here is that some kinds of activity typical of industry do not take place in universities (and conversely), and some kinds of animals, say, in Australia are not found elsewhere. Hence terms are needed, to name and describe activities and animals of such kinds that are not needed elsewhere. Ryle gives a similar example:

> We know that a lot of people are interested in poultry and would not be surprised to find in existence a periodical called "The Poultry World." Here the word "world" is not used as theologians use it. It is a collective noun used to label together all matters pertaining to poultry-keeping. It could be paraphrased by "field" or "sphere of influence" or "province."[9]

In all these cases, a new language or extension of a language is needed to describe special classes of phenomena. The worlds then comprise these special classes or the special concepts invoked in their description.

This sense of "two worlds" is *prima facie* distinct from that in which alternative sets of concepts may be invoked to describe the same phenomena, as in field and particle formulations of a physical theory, functional and causal descriptions of organic processes, or individualist and class descriptions of economic behavior. Here, in contrast to the first cases, the different descriptions may conflict, since they are taken to be of the same class of phenomena. The conflict may itself be described in terms of varying apparent strength, ranging from comparisons of simplicity to clashes of ontology. Thus, to take one of the cited examples, one may argue whether it is simpler to speak of, or whether there exist, electric particles exerting forces as opposed to electric fields with singularities, and similarly with the other examples. For the moment, it is convenient to put such conflicts in more neutral terms as being about the "adequacy" of alternative languages to provide descriptions in some empirical domain (i.e., "world" in the first sense), although on the analysis I assume later, this is equivalent to apparently stronger ontological disputes. But however they are put, such conflicts seem to be indepen-

---

[8] As in the title of Conan Doyle's *The Lost World*, which is a novel partly set in such a region.
[9] Ryle [22], p. 73.

dent of whether either language is also adequate for the description of some other domain.

In fact, the problems raised by these two senses of "two worlds" are not so readily separated, since domains cannot be demarcated independently of languages adequate for describing them. Consider two languages, L_1 apparently adequate to describing the academic structure of Oxbridge, but only of Oxbridge, and L_2, adequate to describing that of all other English universities. To be satisfied that L_2 (which lacks such terms as "college" and "tutor") is also adequate to describing Oxbridge academically is to be satisfied that Oxbridge is not an academic domain, a world, distinct from the rest of the English academic world. And this is to be satisfied that L_1 is inadequate *simpliciter*, since it is admittedly inadequate for other universities, and the adequacy of L_2 shows that there is no basis, independent of L_1, for demarcating an academic domain of Oxbridge, within which alone L_1 could be considered adequate. On the other hand, there is at present no language, L_3, lacking academic concepts, whose adequacy in a domain wider than the academic world would show the latter to be a similarly spurious domain and L_2 to be itself inadequate. It has, of course, been suggested that there is such a language L_3, namely that of industry. For example, Clark Kerr, sometime President of the University of California, makes the suggestion in *The Uses of the University*:

> Basic to this transformation (... now engulfing our [i.e. American] universities ...) is the growth of the "knowledge industry." ... The production, distribution, and consumption of "knowledge" in all its forms is said to account for 29 per cent of gross national product ... and "knowledge production" is growing at about twice the rate of the rest of the economy. What the railroads did for the second half of the last century and the automobile for the first half of this century may be done for the second half of this century by the knowledge industry ... and the university is at the center of the knowledge process.[10]

Much criticism of ex-President Kerr seems to rest on the conviction that this language L_3 is not adequate for the academic world and that turning the University of California into an institution for which L_3 was adequate would render the title "University" inappropriate.

If we turn from universities to the universe the problem becomes at once more interesting and more pressing. There is a view on which, given that the universe comprises one fundamental domain in which all other domains are included, these other domains are spurious, and languages adequate for only some of them are not really adequate

[10] Kerr [8], pp. 87–88.

PHYSICS AND FURNITURE 175

at all. The only adequate language is one adequate for the whole universe, and the only language with such pretensions is that of theoretical physics, L^*. Ryle expresses this view, which he rejects, thus:

> There is nothing that any natural scientist studies of which the truths of physics are not true; and from this it is tempting to infer that the physicist is therefore talking about the cosmos. So, after all, the cosmos must be described only in his terms....[11]

In ontological terms, the claim is that the only things that exist are those over which the individual variables of L^* range. Thus Eddington says: "Modern physics has... assured me that... my second scientific table is the only one which is really there." This is the version of Eddington's view explicitly upheld by such contemporary philosophers as Feyerabend, Sellars, and Maxwell. Feyerabend puts the matter in terms of "concepts" rather than "entities," but the gist is the same:

> ... the "uninstructed layman" does not think of molecules when speaking about the temperature of his milk.... However... a person who has already accepted *and understood* the theory of the molecular constitution of gases, liquids and solids cannot at the same time demand that the premolecular concept of temperature be retained. It is not at all denied by our argument that the "uninstructed layman" may possess a concept of temperature that is very different from the one connected with the molecular theory (after all, some "uninstructed laymen" intelligent clergymen included, still believe in ghosts and in the devil). What is denied is that anybody can consistently continue using this more primitive concept and at the same time believe in the molecular theory.[12]

Sellars puts the point directly in ontological terms:

> According to the view I am proposing, correspondence rules would appear in the material mode as statements to the effect that the objects of the observational framework *do not really exist—there really are no such things*.[13]

Maxwell, who argues in [10] for a modest realist view of theories, has subsequently adopted the "super-realist" view of Feyerabend and Sellars, namely that the entities of physical theory do not merely exist alongside (or inside...) those of everyday speech but displace them; that physical theory shows that the latter do not exist (e.g., [11]).

In contrast to these views we may set those of Ryle [22]. Being

---
[11] Ryle [22], p. 74.
[12] Feyerabend [4], p. 83.
[13] Sellars [24], p. 76.

concerned to preserve the adequacy of everyday language domain, Ryle takes the curious view that scientific languages afford no descriptions at all, and in particular that L^*, which alone claims the universe as its domain, does not do so.

> ... physical theory, while it covers the things that the more special sciences explore and the ordinary observer describes, still does not put up a rival description of them.... It need not be a matter of rival worlds of which one has to be a bubble-world, nor yet a matter of different sectors or provinces of one world, such that what is true of one sector is false of the other.... In the way in which the joiner tells us what a piece of furniture is like and gets his description right or wrong (no matter whether he is talking about its colour, the wood it is made of, its style, carpentry or period), the nuclear physicist does not proffer a competing description, right or wrong, though what he tells us the nuclear physics of covers what the joiner describes.[14]

This could be construed as an extreme instrumentalist view, being applied to *all* general scientific statements, laws as well as theories, and Ryle indeed expresses such a view in *The Concept of Mind*.[15] But if the business of the scientist is as radically different from that of the "ordinary observer" as Ryle suggests, the instruments of science can hardly link everyday observation statements, although they may link scientific observation statements. Now if Ryle's view is correct, all problems of the relation between everyday and scientific descriptions of a domain disappear. This view must therefore be dealt with first, before proposed solutions to these problems can profitably be considered.

Ryle considers the example of an economic theory dealing with buying and selling by individuals.[16] It contains such terms as "the consumer," "the tenant," and "the investor." My brother, a named individual, may of course on occasion be acting as consumer, tenant, or investor, and the theory's predictions about how he will act on such occasions may be true or false. Yet, for Ryle, "in one way the Economist is not talking about my brother," since the theory does not refer to him, he is not named in it; it does not depend on his existence, or "What kind of a man he is. Nothing that the economist says would require to be changed if my brother's character or mode of life changed." This is the important sense in which, Ryle says, the theory does not describe my brother either truly or falsely: "We no longer suppose that the economist is offering a characterization or

---

[14] Ryle [22], p. 80.
[15] Ryle here characterizes law-statements as inference-tickets which "do not state truths or falsehoods of the same type as those asserted by the statements of fact to which they apply." [21], p. 121.
[16] Ryle [22], p. 70.

even a mischaracterization of my brother or of anyone else's brother." Thus this economic pseudo-description is not a rival to the real everyday description of my brother.

The sense in which economic theory is "about" my brother and does "describe" him is, for Ryle, the trivial (or at least irrelevant) sense in which it is about, or describes, anyone who is a consumer. "In another way the economist certainly is talking about my brother, since he is talking about anyone, whoever he may be and whatever he may be like, who makes purchases, invests his savings, or earns a wage or salary." The economist may say: "The consumer is ... ," or "the consumer does ... ," which is, of course, just to say: "All consumers are ... ," or "All consumers do. . . ." So we may say that these statements of economic theory are true or false of my brother *in his rôle of consumer*; they describe him *as a consumer*, but they do not describe *him*.

This distinction, between the real everyday description of my brother and the pseudo-description of him by economic theory merely in his rôle of consumer, is both central to Ryle's thesis and entirely spurious. Ryle gives no clear examples of everyday description in the sense which economic theory is unfitted to provide. He merely observes that the economist does not need to know "that I have a brother, or what kind of a man he is." The first part of this, the bare existential assertion, is not a promising candidate for a distinctively everyday description. Someone giving an everyday characterization of brothers in general does not need to know that I have one. Everyday descriptions of brothers doubtless presuppose that there are brothers, but then so do economic descriptions of consumers presuppose that there are consumers—otherwise the theory would be of a mythology, not a science. And those who accept and those who reject Russellian descriptions agree that an economist and a plain man, setting out specifically to describe *my* brother, will equally fail to make true statements if I have no brother.[17]

The spuriousness is perhaps less obvious of a distinction between everyday "real," and economic pseudo-, descriptions that is based on Ryle's concept of "kinds" of men. The idea here is that the everyday description shows what "kind" of a man my brother is, as the economic description does not. Now the dispositional qualities that mark a "kind" of man should presumably be fairly permanent. But the most obvious examples are certainly not immutable: a *tolerant* man may become dogmatic, a *thin* man may grow fat. This sort of change must evidently be distinguished from the sort of change that

[17] E.g., Quine [18] and [19], and Strawson [26].

constitutes starting to display a disposition. A consumer need not consume all the time, but the disposition he has to consume in a certain way is present whether he is actively consuming or not. My brother does not become a certain kind of consumer simply by starting to consume, and cease to be of such a kind when he stops consuming. Changes in his economic *activities* are clearly distinct from changes (if any) in his economic *dispositions*—and it is the latter that economic theory describes.

Perhaps an economic disposition is not permanent enough for Ryle to characterize a "kind" of man, or perhaps it is not important enough. But the same may be as true of an everyday description and, conversely, economic character may be stable and important even in a philosopher's brother. A financially *prudent* (or careful, or mean) brother surely instances a kind of man as much as a *brave* brother. A prudential disposition may not show itself except in economic situations that may be rare (though perhaps not rarer than those which call for displays of bravery), but that observation is not to the point. Popular superstition and the paradoxes of material implication notwithstanding, a man does not influence his weight by declining to weigh himself (except perhaps causally).

Ryle's other examples, of an accountant's view of a College or library,[18] equally fail to show that such technical statements do not provide descriptions. The more interesting claim that he also makes in these cases is that such statements do not constitute *complete* descriptions of these institutions. Ryle makes this claim too glibly; the matter is more complex than he suggests. Certainly, from the fact (if it were so) that a college or library account said something about every college activity or every library book, it would not follow that it said everything about any (let alone every) activity or book. If we call such accounts "*comprehensive*" in the sense that they cover all items of which an accountant's description can be given, we may say that a comprehensive economic description of a domain need not contain a *complete* description of any item in that domain. Similarly, a comprehensive chemical description of a set of things will not contain a complete description of any member of the set, since every member will also have non-chemical (e.g., physical) properties. But we would think it arrogant of a physicist (whose everyday language is that of physics) to insist that therefore the chemical description of such a thing is not a real (i.e., physical) description at all. A description does not have to be complete to be a description.

However, Ryle's accountant could make a more interesting claim

[18] Ryle [22], pp. 75–78.

for completeness than the one Ryle lets him make. He could claim that the various other descriptions of college activity or library books could be reduced to (or replaced by, or translated into) financial descriptions. For example, he could claim that the value a student placed on an activity or book could in principle be described completely in financial terms. This is the claim analogous to that of the theoretical physicist: not the claim that descriptions in languages other than L^* are not given (which is clearly and trivially false), but the claim that all such descriptions could in principle be translated into descriptions in terms of L^*. Such a claim may be as implausible as the corresponding accountant's claim, but neither is refuted by pointing to the obvious and admitted fact that descriptions, within their various domains, are given in terms of other languages. Yet Ryle links his denial, of the library accountant's supposed claim, to the facts of usage thus: "The student's information about the books is greatly unlike the accountant's, and neither is it deducible from the accountant's information, nor *vice versa*...."[19] This admittedly does not have the explicit form of an inference, but since Ryle gives no other grounds for denying deducibility, I think it must be read as such.

But, if completeness is taken as I suggest, on what grounds are claims about the unique adequacy of L^* to be assessed? What is being claimed for L^* if not, what is clearly false, that the language is used for every domain of things? Presumably that in principle L^* *could* be so used, supplemented with explicit definitions of complex terms to replace such everyday equivalents as "table" and "man." But it is far from clear what principle it is that properly assures us of this remote possibility, and what the grounds are for accepting it. The principle seems to be an ontological one, depending on a realist view of the theories expressed in L^*. And before this can be further discussed, some clarification of the notion of an "ontological principle," of what a realist view of theories commits one to, is needed.

For present purposes, I follow Quine in his analysis of existence. For a *named* individual, class, or attribute to exist is to be *the* value of a variable;[20] for an individual, class, or attribute of some *kind* to exist is to be *a* value of a variable.[21] But whether or not this particular analysis is accepted, existence, like truth, seems too fundamental a concept for the term "existence" and its synonyms to admit of different senses, such that an item could both exist in one sense and

---

[19] *Ibid.*, p. 78.
[20] Quine [17], p. 50.
[21] Quine [19], p. 224.

not exist in another. We may admit different *kinds* of existents, or entities, where items are classified by the methods used to establish their existence. For example, I reserve the term "thing" for items capable of spatio-temporal location, which are the concern of science, as opposed to such items as numbers. Similarly, we may admit different *kinds* of truths, where true statements are classified by the methods used to establish their truth. Thus we may distinguish mathematical from physical entities just as we may distinguish mathematical from scientific truths. But no more than the latter distinction requires distinct mathematical and empirical senses of "true" does the former require distinct mathematical and empirical senses of "exist."

This point needs to be made because it is tempting to try and evade the problem by saying that everyday and theoretical entities, like scientific and mathematical entities, can both exist, but in different "ways" or senses; just as God is sometimes said to exist in a different way from the rest of us. One is reminded of Mill's famous outburst against the convenient multiplication of senses in theology:

> Language has no meaning for the words Just, Merciful, Benevolent, save that in which we predicate them of our fellow-creatures; and unless that is what we intend to express by them, we have no business to employ the words. If in affirming them of God we do not mean to affirm these very qualities... we are neither philosophically nor morally entitled to affirm them at all.... To say that God's goodness may be different in kind from man's goodness, what is it but saying, with a slight change of phraseology, that God may possibly not be good?[22]

And in this respect at least the same is true of the existence of theoretical entities. What Russell calls "a robust sense of reality" is as necessary in philosophy of science as in logic, and he who pretends that electrons, say, have "another kind of reality" does as much "disservice to thought" as he who pretends that "Hamlet has another kind of reality"[23]—more indeed, since physical theory, unlike *Hamlet*, is not advanced as a piece of fiction.

In particular, it is regrettable that Nagel[24] should have taken his elaborate discussion of the different criteria for what he calls "physical reality" or "physical existence"[25] as establishing different "senses of 'real' or 'exist' that can be distinguished in discussions about the reality of scientific objects."[26] Of course, it establishes nothing of the

[22] Mill [13], pp. 122–123.
[23] Russell [20], p. 170.
[24] Nagel [14], pp. 146–151.
[25] *Ibid.*, p. 146.
[26] *Ibid.*, p. 151.

sort, as Maxwell rightly observes;[27] what it *does* establish is (a) that the existence of different "scientific objects" is established in different ways, and (b) that there are disputes about what the ontological commitments of particular theories are. These latter disputes, which are the subject of this paper, would not of course be resolved merely by accepting such an analysis of the *one* sense of "existence" as that of Quine's. An instrumentalist who will not admit the existence of electrons would then not admit a formalization of physical theory in which a variable ranges over electrons—except as a convenient "eliminable shorthand." But then, as Quine observes in a different context, the onus is on the instrumentalist to "devise contextual definitions explaining quantification with respect to [these] alleged entities."[28] Otherwise, "he will perhaps still plead that his apparent ... entities are merely convenient fictions; but this plea is no more than an incantation, a crossing of the fingers, so long as the required contextual definitions are not forthcoming."

The point of invoking the Quinean analysis is that it shows in what terms ontological disputes may be resolved. It is characteristic of a theory not merely to make new statements but also to provide "a new language or extension of a language" in which to make them. Such is the language L^* in which physical theory T is formulated. If alternative formulations, with variables ranging over different things, are accepted as equivalent, i.e., as of the same theory, T, then an instrumentalist view is being taken at least of those things not common to both formulations. Conversely, a realist view must distinguish these formulations as distinct theories, since they carry distinct ontological commitments. To take a recent example, Hoyle's objection to a field formulation of relativity, because it permits a solution in a one-body universe, must be taken by a Quinean to reflect a realist view of the gravitational field.

Thus the ontological commitment of those advocating a theory is shown in the restrictions they impose on languages in which it can be stated. Fortunately, the detailed analysis of such commitment is beyond the scope of this paper, since it is common ground that, however weak the restrictions are, everyday language cannot satisfy them, i.e., that everyday language is inadequate for the domain of fundamental physics. There is therefore no doubt, on this analysis, that physical theory T assumes the existence of *some* things other than those assumed in everyday language. Disputes about just *what*

---

[27] Maxwell [10], pp. 20–21.
[28] Quine [17], p. 51.

things are assumed—fields or particles, for example—are not to the present purpose.

What is to the purpose is the "super-realist" claim that the ontology of whatever L^* is most adequate to physical theory T must be *complete* since L^* alone has the universe as its domain and T is the best supported theory of the universe. Let us, for convenience, call the things ranged over by variables in T, whatever they are, "basic things." Then other things, such as tables, that seem to exist independently in the physical world do not really do so; there are simply in their place (so to speak) suitably arranged collections of basic things. This is the super-realists' ontological principle. But let us press the question: what are, or could be, the grounds for thinking that this ontological principle is true? It seems only to entail that all statements ostensibly about tables, etc., can be translated into statements about basic things, but not conversely; and this claim begs the very question at issue, since it is admitted that such translations are not made, and as a matter of fact cannot be made with our present mathematical expertize. If it is insisted that such translation could be carried out in principle, we observe that the possibility of translation *is* the principle that is supposed to establish the unique adequacy of L^*. If this possibility is itself only established by assuming the unique adequacy of L^*, the argument is viciously circular.

In fact, I shall argue that the translatability claim is false and that, when it is analyzed a little more closely, common knowledge shows it to be false. The point of the claim seems to be that a statement about everyday things is either equivalent to some theorem of T or is false. No doubt, if current theoretical physics is true, this is true. But this is just what we don't know in any sense strong enough to support the conclusion. The translatability claim is, after all, not made for the "ultimate" physical theory (if any) of which we know nothing—except that it is true. The claim is made for the *current* theory T, of which all we know, at best, is that it is well supported by empirical evidence.

The translatability claim must therefore be that *T is better supported than any of the apparently independent statements that are to be translated into L^**. Only if this is so can we reasonably claim to know that any such statement must either be equivalent to a theorem of T or be false. Now if this were so, then suppose that from T, by dint of Eddington's "remorseless logic," we can extract as a theorem a statement S equivalent to "mahogany is blue." Until derivations of this complexity (in terms of L^*) have been carried out, which they haven't, we don't know that S is *not* a theorem. And if S *is* a theorem of T, the translatability claim entails that our pre-theoretical preference for $\sim S$ must be discarded. It entails that *any future conflict between a theorem of T and an accepted everyday statement must be resolved by discarding the everyday statement as false.* But a "remorseless logic" that leads to this conclusion would do better to show remorse. Experiments and observations peculiar to T, i.e., such that their results are describable only in L^*, are very few and doubtful compared with those that also lend support to statements in less comprehensive languages (such as, for example, that of atomic physics). The vast bulk of the evidence for T is also, and more directly, evidence for statements in other languages than L^*. Hence the constraint is on T, that *its* statements about collections of fundamental particles be either equivalent to accepted statements about the color of mahogany or be rejected as false, not *vice versa*.

There are, of course, borderline cases where the evidence for T and statements in other languages is on a par, and T reconstructs previously accepted statements. But rather too much has been made of these cases. Thus Popper:

> It is well known that Newton's dynamics achieved a unification of Galileo's terrestial and Kepler's celestial physics. It is often said that Newton's dynamics can be induced from Galileo's and Kepler's laws, and it has even been asserted that it can be strictly deduced from them. But this is not so: from a logical point of view, Newton's theory, strictly speaking, contradicts both Galileo's and Kepler's (although these latter theories can of course be obtained as approximations, once we have Newton's theory to work with).[29]

As a point against the necessity of an inductive *method* in science, this is well taken: that Newton could not have argued inductively in this case certainly shows that scientists do not always do so. It may be that they never do so, as Popper asserts (e.g., in [16]), or only do so in periods of "normal" science in between paradigm changes.[30] But whatever the fact is (see, e.g., Achinstein [1]) it is merely psychological, about how scientists actually make progress; *pace* Feyerabend,[31] it is not a *logical* fact, that "the contents of a whole theory (and thereby again the meaning of the descriptive terms which it contains) depends on ... the set of all the alternatives which are being discussed at a given time."

The admitted existence of theoretical controversy and the modification by theories of laws which they have successfully explained have

---

[29] Popper [15], pp. 29–30.
[30] See Kuhn [9].
[31] Feyerabend [4], p. 68.

been much exaggerated in importance.[32] That "one and the same set of observational data is compatible with very different and mutually inconsistent theories"[33] is not a startling new discovery; it is a well-known elementary logical fact. Neither it nor the other facts cited go any way to support the super-realists' ontological principle. If to be is to be a value of a variable, then what we can assume exists is what is quantified over in well supported law-like statements, and T is *not* better supported than more mundane laws. Ontology is dependent upon epistemology, since what we can know to exist is merely a part of what we can know.

So a reasonable realist view of theories takes them on their first acceptance as *adding* to the stock of known things. As their continued success leads scientists in fact to treat the statements they explain as replaceable by those of the theory, so and no further may the pre-theoretical things be taken to be replaced by those of the theory. But success in explaining what is not yet replaced is a precondition of such ontological replacement; it does not comprise it. Nagel is right in insisting on an *initial* meaning invariance of accepted laws and theories under further theoretical explanation:

> Despite what appears to be the complete absorption of an experimental law into a given theory, so that the special technical language of the theory may even be employed in stating the law, the law must be intelligible (and must be capable of being established) without reference to the meanings associated with it because of its being explained by that theory. Indeed, were this not the case for the laws which a given theory purportedly explains, there would be nothing for the theory to explain.[34]

And similarly for the terms in a theory whose success is to be explained by its reduction to a theory of another science:

> ... expressions distinctive of a given science (such as the word "temperature" as employed in the science of heat) are intelligible in terms of the rules or habits or usage in that branch of study, they must be understood in the senses associated with them in that branch *whether or not the science has been reduced to some other discipline*.[35]

Nagel is wrong only in implying here that successful explanation or reduction may not be *followed* by a shift in meanings. The point is that once a theory T_1 is successfully reduced to another, T_2, the very success of T_2 in its wider domain may lead us to abandon the narrower domain (of terms, and hence of things) within which T_1 was

---

[32] See Mellor [12], and on "meaning-variance" between terms employed in conflicting theories, see Fine [6] and Hesse [7].
[33] Feyerabend [4], p. 48.
[34] Nagel [14], p. 87.
[35] *Ibid.*, p. 352. My italics.

admittedly adequate. We may still, of course, retain the terms of T_1 for their convenience within its domain, without ontological commitment. The things they purport to refer to *are* then regarded as "convenient fictions." But until such a replacement, subsequent to successful reduction, of T_1 has in fact taken place, no ontological savings have been made; and until then no principle can guarantee that such savings will be made. To assert otherwise is to issue a blank check on the future success of current theory. Such checks having always bounced in the past, there is no good inductive reason to accept one now.

In fact, the ontological replacement even attempted by current physical theory is very modest, extending no further than things of molecular dimensions. The domain of L^* is not in fact universal: it is simply the domain of the very small. No doubt the very small can be found everywhere, but so can the very large. If it is true that inspecting common objects more closely reveals that their parts are atoms, it is equally true that inspecting them more distantly reveals that they in turn are parts of galaxies. The preference for explanation in terms of the very small is, as Schlesinger has observed,[36] a logically unwarranted prejudice (for those who share it, a regulative principle . . .), namely that of "micro-reduction." A similar principle of "macro-reduction"[37] might equally be held, as a determination to explain everything functionally, in terms of larger wholes of which it is a part. Indeed, Mach's Principle, that the inertial properties of a thing should be taken to be a function of the positions of all other things, is an illustration of just such an attitude. Either principle may be justified pragmatically by the scientific discoveries to which it gives rise. Neither is justified ontologically until either all talk of electrons can be not merely reduced to, but replaced by, talk of galaxies, or conversely. Until then, the universe, so far as we have reason to believe, contains as independent entities both galaxies and electrons—*and* tables and men. And it is more likely to continue to be reasonable to believe in the independent existence of tables and men than in the existence of electrons and galaxies, well supported as theories referring to the latter are. No conceivable theories either of the very large or very small are likely to carry their replacement of current things as far as tables and men; and any that do will certainly have replaced electrons or galaxies first.

Pembroke College, Cambridge

[36] Schlesinger [23], p. 46.
[37] *Ibid.*, p. 56.

References

[1] Peter Achinstein, Review of Popper's *Conjectures and Refutations*, British Journal for the Philosophy of Science, vol. 19 (1968), pp. 159–168.

[2] Rudolph Carnap, "P. F. Strawson on Linguistic Naturalism" in P. A. Schilpp (ed.), *The Philosophy of Rudolph Carnap* (LaSalle, Illinois, 1963), pp. 933–940.

[3] A. S. Eddington, *The Nature of the Physical World* (London, 1929), Introduction.

[4] P. K. Feyerabend, "Explanation, Reduction and Empiricism" in H. Feigl and G. Maxwell (eds.), *Minnesota Studies in the Philosophy of Science*, vol. III (Minneapolis, 1962), pp. 28–97.

[5] —— "The Structure of Science," British Journal for the Philosophy of Science, vol. 17 (1966), pp. 237–249.

[6] A. I. Fine, "Consistency, Derivability and Scientific Change," The Journal of Philosophy, vol. 64 (1967), pp. 231–240.

[7] M. B. Hesse, "Fine's Criteria of Meaning Change," The Journal of Philosophy, vol. 65 (1968), pp. 46–52.

[8] Clark Kerr, *The Uses of the University* (Cambridge, Mass., 1963), ch. 3.

[9] T. S. Kuhn, *The Structure of Scientific Revolutions* (Chicago, 1962).

[10] Grover Maxwell, "The Ontological Status of Theoretical Entities" in H. Feigl and G. Maxwell (eds.), *Minnesota Studies in the Philosophy of Science*, vol. III (Minneapolis, 1962), pp. 3–27.

[11] —— "Scientific Methodology and the Causal Theory of Perception" in I. Lakatos and A. Musgrave (eds.), *Problems in the Philosophy of Science* (Amsterdam, 1968), pp. 148–160.

[12] D. H. Mellor, "Experimental Error and Deducibility," Philosophy of Science, vol. 32 (1965), pp. 105–122.

[13] J. S. Mill, *An Examination of Sir William Hamilton's Philosophy* (London, 1867).

[14] Ernest Nagel, *The Structure of Science* (New York, 1961).

[15] Karl R. Popper, "The Aim of Science," Ratio, vol. 1 (1957), pp. 24–35.

[16] —— "Philosophy of Science: a Personal Report" in C. A. Mace (ed.), *British Philosophy in the Mid-Century* (London, 1957), pp. 155–191.

[17] W. V. O. Quine, "Designation and Existence" in H. Feigl and W. Sellars (eds.), *Readings in Philosophical Analysis* (New York, 1949), pp. 44–51.

[18] —— *From a Logical Point of View* (Cambridge, Mass., 1953), ch. 1.

[19] —— *Methods of Logic*, 2nd ed. (London, 1962), Pt. 4.

[20] Bertrand Russell, *Introduction to Mathematical Philosophy* (London, 1919), ch. 16.

[21] Gilbert Ryle, *The Concept of Mind* (London, 1949), ch. 5.

[22] —— *Dilemmas* (Cambridge, 1954), chs. 5–6.

[23] George Schlesinger, *Method in the Physical Sciences* (London, 1963), ch. 2.

[24] Wilfrid Sellars, "The Language of Theories" in H. Feigl and G. Maxwell (eds.), *Current Issues in the Philosophy of Science* (New York, 1961), pp. 57–77.

[25] L. S. Stebbing, *Philosophy and the Physicists* (London, 1938), chs. 3–4.

[26] P. F. Strawson, *Introduction to Logical Theory* (London, 1952), ch. 6, §7.

[27] —— "Carnap's Views on Constructed Systems versus Natural Languages in Analytic Philosophy," *The Philosophy of Rudolph Carnap*, ed. by P. A. Schilpp (LaSalle, Illinois, 1963), pp. 503-518.

X

Religion, Science, and the Extraordinary

MICHAEL ANTHONY SLOTE

IN the present paper I shall introduce a type of argument for the existence of some sort of God or higher being which, despite its striking simplicity, has never (as far as I can tell) been explicitly promulgated by any philosopher or religious thinker. I shall claim that this argument has at least *some plausibility*, due to the fact that it is based on a very general principle whose validity seems to be presupposed in much thinking both inside and outside the area of religion.

I

The notion of the extraordinary (and also the related notions of the remarkable, the amazing, and the wondrous) plays, I believe, an important role in the attempts of men to understand the world. Consider, for example, an idealized primitive man confronted for the first time with the phenomenon of thunder (or thunder-cum-lightning). This phenomenon will be different from anything else he experiences in everyday life, and will constitute something amazing and extraordinary for him. I want to suggest that if the primitive man seeks to explain the thunder (or thunder-cum-lightning) in terms, e.g., of a god beating a great drum, or a god striking huge flints—which is the sort of explanation often given by primitive people(s) of such phenomena—what he is doing is explaining what strikes him as remarkable or extraordinary in what seems to him the most appropriate sort of way, namely, in terms of something equally extraordinary, that is, in terms of the actions of an extraordinary being[1] with respect to extraordinary objects.[2]

---

[1] After all, such a god is much larger and more powerful than the conscious beings he confronts in everyday life.

[2] In his *An Introduction to the History of Religion* (London, 1896), F. B. Jevons explains a good deal (but not all) of primitive religious belief in this way, i.e., as a response to certain awe-inspiring or striking events, like thunder and lightning (pp. 19ff.).

Consider the following epistemic principle that seems to be presupposed in the kind of thinking by which a primitive man might, in the manner just described, come to believe that thunder was caused by the actions of some higher being:

> If x is an extraordinary (or amazing, or remarkable) observed, or experienced phenomenon, one has some reason to think (and it is, *ceteris paribus*, reasonable to think) that x has been brought about by (and is thus to be explained in terms of) something extraordinary (or amazing, or remarkable) that is unobserved or that "lies behind" x.

This principle I call the *Principle of Extraordinary Explanation* (PEX, for short, hereafter). It is not only presupposed in some primitive thought, I believe, but also in the relatively sophisticated thinking of some modern philosopers and scientists, and in religious thought in general, though never in an explicit way. Something like this principle seems, for instance, to be behind some of the things Schopenhauer says about music in *The World as Will and Idea*. According to Schopenhauer, everything that the senses perceive is an objectification of the Will achieved by means of ("filtered through") certain Ideas, with the exception of music, which is a direct and unmediated objectification of the Will. He says:

> ... music ... since it passes over the Ideas, is entirely independent of the phenomenal world, ignores it altogether, could to a certain extent exist if there was [sic] no world at all, which cannot be said of the other arts.... Music is thus by no means like the other arts, the copy of the Ideas, but the *copy of the will itself*, whose objectivity the Ideas are. This is why the effect of music is so much more powerful and penetrating than that of the other arts, for they speak only of shadows, but it speaks of the thing itself.[3]

This passage suggests strongly that Schopenhauer thought that only a major break or discontinuity in his metaphysical system could explain the extraordinary force and beauty of music. (Soon after the above passage, he talks of the "unutterable depth of all music.")[4] Only by positing a unique direct objectification of the Will in human experience could the remarkable nature of musical experience be explained. And from the point of view of Schopenhauer's system, it is indeed extraordinary that the Will should reveal itself in such an immediate manner in this single case. Thus it seems that Schopenhauer in positing something extraordinary or remarkableto explain something he finds extraordinary or remarkable in experience is implicitly using or presupposing the PEX.

[3] *The Philosophy of Schopenhauer*, ed. by I. Edman (New York, 1928), p. 201.
[4] *Ibid.*, p. 209.

Perhaps the area where we can best see the PEX at work, however, is not philosophy, but religion. A great deal of religious conviction can, I think, be plausibly understood as resulting or receiving sustenance from implicit application of this principle. A dream, for example, can sometimes occur with such force and vivacity as to persuade him who has had it that it was more than just another (crazy) dream. If one has a dream vision of the Lord of Hosts, or of the Virgin Mary, one may wonder whether one has been drugged, or is going mad, or one may think that one's dream was the result of mere wish-fulfillment. But some people's lives are transformed by just these sorts of experiences. They come away with a sense of wonder and amazement, feeling that what they have experienced is so extraordinary and remarkable that it cannot be "explained away,"[5] as one would tend to do with most dreams, as being due entirely to one's subconscious desires and fears, to drugs, or to the overtime workings of one's imagination. And so they will think that their dream has come from God himself or the Virgin herself,[6] making use in an implicit way of the PEX.[7] (If God and the Virgin exist, they are, presumably, extraordinary beings.)

People who have not had any special dream experiences also seem to make use of the PEX. Many people, for example, are struck and moved by the immense order and beauty of the world, and this may cause them to believe or reinforce their belief that this order and beauty, this extraordinary network of interlocking intricacies, calls for explanation in terms of some sort of extraordinary being behind nature as a whole. Or, in a slightly different vein, they may feel that the awesome fact that there exists something (contingent), rather than nothing (contingent) at all, in the universe calls for the same sort of explanation.[8] Others find particular things in the world or facts about the world, rather than the world as a whole, remarkable. For some it is the extraordinary capacities of the human mind that call for

[5] A man who knows he took an hallucinogen before having a certain dream experience might agree that the hallucinogen had *something* to do with his experience; but he might well want to say that God *also* had something to do with it.

[6] I have not been able to find any written descriptions of such dream experiences, but an actual occurrence of such an experience has been reported to me by S. Silberblatt, a student at Columbia College. He had a dream about God that he considered more remarkable than anything he had experienced before, and according to him, his current religious faith stems from that dream.

[7] But not just the PEX, which does not permit explanation of something extraordinary in terms of any particular extraordinary being, like God. More on this below.

[8] See A. Heschel, *Man is Not Alone* (New York, 1951), p. 12; also his *God in Search of Man* (New York, 1955), pp. 45ff.

explanation in terms of the PEX.[9] For others it might be the inexorableness of natural law or the intricacy of a leaf or rock crystal that called for such explanation. But in any case, whether one is going from the extraordinary character of the world as a whole or of particular facts or things, those who postulate the existence of some extraordinary higher or deeper force or being behind things along these lines seem to be presupposing the validity of something like the PEX.

II

The PEX can only be used in cases where one believes something to be amazing, remarkable, or extraordinary.[10] But what is it for something to be amazing, remarkable, or extraordinary? And do we ever have (objectively) good reasons for calling something extraordinary, etc.? This latter question arises because judgments of extraordinariness or remarkableness seem very much like value judgments inasmuch as they express individual attitude and preference. So just as some philosophers have been Emotivists or Subjectivists about value words, so too would it be possible to maintain an Emotivist or Subjectivist position with respect to words like "extraordinary," "amazing," etc. One might claim, in other words, that to say that something is extraordinary, etc., is just to evince, or describe, one's emotion of awe or wonder at it. However, both Emotivism and Subjectivism about value terms are on the wane today, as a result of various recent and well-known criticisms that have been levelled against these positions. And these criticisms, I think, will serve equally well to undermine Emotivism and Subjectivism with regard to such words as "extraordinary," "amazing," etc.

Perhaps the most enlightening and thoroughgoing treatment of terms like these that has been undertaken in recent years occurs in Patrick Nowell-Smith's *Ethics*.[11] According to Nowell-Smith, such words as "extraordinary" and "amazing" belong to a class of adjectives that he calls "A-words" and that includes words like "sublime" and "disgusting" as well. He says that although there is an emotive component to A-words, there are, nonetheless, standards for judging whether something is in fact sublime or amazing or

[9] Heschel, *Man is Not Alone, op. cit.*, p. 14.

[10] And not even in all such cases. According to the PEX, extraordinary *observed or experienced* phenomena call for explanation in terms of something extraordinary. This *may* also be true of extraordinary *unobserved* entities, but I have seen no evidence that this is so. Thus the PEX, as stated, is not committed to any infinite regresses of extraordinary explanantia.

[11] London, 1954, pp. 70-91.

disgusting. Such judgments are, of course, often the subject of serious dispute, but they are still descriptive and statement-making and can be argued about rationally. I agree with much of what Nowell-Smith is saying in this matter. Statements about what is or is not amazing or remarkable or extraordinary assert something and can be supported by good reasons. If a man dreams that his wife and children have just been killed in the crash of a plane whose departure was delayed six hours and which contained Bolivian soccer players and was flown by a Canadian pilot, then if in fact all these things are true, and the man had no way of knowing any of them, his dream is remarkable and amazing, because it involves a remarkable and amazing coincidence. One might dispute that the dream was mere coincidence, and claim the man had been secretly informed, but if one grants the facts claimed, one has to admit that a remarkable, extraordinary thing occurred.

Of course, what is or is not amazing cannot be so easily ascertained in all cases. Consider someone who thinks that the intricacy, order, and beauty of the universe are extraordinary and who consequently feels that there must be something very remarkable that explains all that intricacy, etc. His train of thought relies on the assumption that certain aspects of the world are extraordinary. Now some of us are more impressed or moved by those aspects of the world than others, and are thus more ready to see them as extraordinary than others. And, of course, it is impossible to *prove* either that the world is or that the world is not remarkable and extraordinary in certain of its aspects. Since John Wisdom's article "Gods" and others of his writings,[12] philosophers have become accustomed to the thought that disputes over questions like whether the order and beauty of the world are remarkable depend at least in part on different ways of seeing things, on differences of sensitivity and upbringing, etc. Although I myself believe (the order and beauty of) the world to be remarkable and extraordinary, worthy of wonder and awe, and can point out reasons[13]

---

[12] Compiled in *Philosophy and Psychoanalysis* (Oxford, 1957).

[13] There are several sorts of reasons that clearly count in favor of something's being extraordinary or remarkable. If we learn that thing x (a) is statistically rare; (b) cannot be reproduced by us; or (c) possesses some valued, or disvalued, attribute to a far greater degree than we do, we have, I think, discovered some reason for thinking x extraordinary or remarkable. Certain diseases, meetings of people in far-off places, and body traits exemplify (a); the intricacy of a rock crystal or leaf exemplifies (b); and God (if he exists) and heroes and saints exemplify (c). Even though there are different sorts of reasons for remarkableness and extraordinariness, we need not conclude that "extraordinary" and "remarkable" are ambiguous terms. The case is, I think, rather like that of "morally wrong." There are many different sorts of reasons for calling something morally wrong, all in the same sense of that expression.

for thinking so, I cannot *prove* that it is in any final, decisive way.

Of course, one may argue for the existence of something extraordinary behind things on the basis of certain remarkable dreams, rather than on the basis of the remarkableness of the order and beauty of the world. But the remarkableness of a dream may also come into dispute, in great part because dreams are private: are had by one person only. Thus if x has a dream and tells y about it, it might be true that if y had had an exactly similar experience, he too would have found it remarkable. But if he has not had such an experience himself, then he may not be able to see or understand the extraordinariness of what has happened to x, or thus to see or understand why x thinks his dream must have come from some higher being. Furthermore, it may just be impossible for x to convince y of the extraordinariness of his dream. And this may not be because y is unintelligent or distrustful of x, but because no amount of description on x's part may be able to convey the power and vividness of x's actual experience, i.e., convey the very factors that were found by x to be extraordinary and remarkable about his dream. It may well be, then, that x has *excellent reason* to believe he has experienced something extraordinary, and yet is unable to convince (certain) others of this. Perhaps, then, there is some truth to the claim often made by theologians and religious people that no one who has not himself had a religious experience has any right or good reason to deny the truth of those religious beliefs that emerge from or depend on such experiences.[14] For if the PEX is valid, someone who has an extraordinary dream has some reason to believe in some sort of remarkable being or entity as its source; and since someone who has never had an extraordinary experience of that sort may not be able to grasp or see the remarkableness of that dream, he will simply fail to have or to appreciate the reason possessed by the dreamer for belief in the existence of some sort of remarkable being behind his dream.

In terms of the PEX we can explain why some people on the basis of experiences, events, or things they find extraordinary come to believe that something extraordinary is behind, is the cause of, those experiences, etc. But we cannot in terms of the PEX alone explain why they so often pick out one or another *particular kind* of extraordinary being or thing as the source or cause of that in their experience which they find extraordinary. One who has a dream about the Virgin, for example, is entitled by the PEX only to conclude

[14] See A. Flew and A. MacIntyre (eds.), *New Essays in Philosophical Theology* (London, 1956), pp. 76ff., 81ff.

that (there is some reason to think that) some sort of extraordinary being has caused it. But that principle no more entitles him to say that his dream comes from the Virgin than that it comes from a green dragon or from Satan. Why, then, do people so frequently go beyond the PEX and posit particular kinds of entities as the extraordinary causes of the extraordinary things they have witnessed?

The answer, perhaps, is that if an event is extraordinary, there is a tendency for people to postulate as the extraordinary cause of it that (sort of) extraordinary explanans which of all possible (extraordinary) explanantia of the phenomenon seems the most likely to be the cause of the phenomenon, given the other beliefs that they have. Thus a good Catholic will say that his extraordinary dream of the Virgin talking to him and telling him to sin no more was sent by the Virgin herself, not by a green dragon, because he already believes in the Virgin and in her occasional appearance to ordinary mortals and also believes that there are no green dragons. And he will believe that the cause of the dream is the Virgin and not the Devil, even though he believes in the existence of both, if his experience is a moving and radiant one and if in it the Virgin tells him to do things he knows are good and if as a result of having the dream he is happier and morally more upright and more zealous religiously. For he believes that the Devil is not interested in making people happier, more upright, or more religiously zealous. But if the same man has a dream of the Virgin in which she tells him to do things he thinks are evil or appears in a vulgar way or acts in an unholy manner, then he might not be so quick to assume that it was the Virgin, and not the Devil that was responsible for the dream. For he believes that the Virgin would not act that way, but that the Devil might try to make it seem as if she would.

I should like, then, tentatively to propose the following *Principle of the Most Appropriate Extraordinary Explanation* (henceforth, PMAEX):

> If x is an extraordinary observed or experienced phenomenon, one has some reason to think (and it is, *ceteris paribus*, reasonable to think) that it has been brought about by the particular (kind of) thing or being y, if on one's other beliefs, y is the most likely extraordinary explanans of x and *if those other beliefs are reasonable*.

In terms of this principle we can understand why a Catholic might hold that his dream of the Virgin had come from the Virgin, for that explanation of his dream is the explanation positing an extraordinary explanans that best fits in with his other beliefs, and he, of course, believes that his other beliefs are reasonable and justified. But those other beliefs include belief in the Virgin Birth of Jesus, the belief that the Virgin exists somewhere, and the belief that the Virgin can appear to ordinary mortals, etc. And to the extent that a good Catholic is not rationally justified or reasonable in maintaining these beliefs, the PMAEX does not grant him reason to think that his dream came from the Virgin.[15] And, of course, even if those other beliefs are not justified, the PEX will at least grant him reason to believe that *some* sort of extraordinary force, being, or thing has brought about his dream.

III

Until now we have been assuming the validity of the PEX and the PMAEX. Perhaps the strongest thing that can be said in their behalf is that they are principles in accordance with which people do implicitly argue. We have already discussed several examples of the use of the PEX and shall be discussing one or two more examples below. Unfortunately, there seems to be no noncircular way of proving the PEX, or the PMAEX. But then it is well-known that there is no way of noncircularly proving the validity of inductive inferences, and yet we still believe that some such inferences are valid. Perhaps, as Nelson Goodman would say, the most that can be said for the validity of some principle of inductive inference is that we want to make particular inferences in accordance with it and find the making of such inferences indispensable in our thinking.[16] And perhaps the same sort of thing is all that can or need be said in favor of the PEX or the PMAEX. However, someone of a scientific or anti-metaphysical bent might at this point object that science requires the acceptance of various principles of inductive inference,

---

[15] However, we should not assume that just because we rationalistic philosophers are not rationally justified in believing in the Virgin, etc., good Catholics are not justified in having such beliefs. A good Catholic may have inductive reason to believe in the Virgin, etc., on the authority of his priest, who he knows is reliable in what he says outside the area of religion and much smarter than he. And if he lives in a backward area, he may not know that morally good, deep-thinking people sometimes do not believe in the Virgin, etc. Given what *we* know, however, *we* are not justified in believing in the Virgin, etc., and cannot have the reason or rational justification for believing in the Virgin, etc., that the man in question has. So perhaps that man would have reason to explain some dream he had of the Virgin in terms of the Virgin, even though we, if we experienced the same sort of dream, would not.

[16] *Fact, Fiction, and Forecast* (Indianapolis, 1965), p. 63. For a defense of enumerative induction against some of its recent critics see my "Induction and Other Minds," *The Review of Metaphysics*, vol. 20 (1966), esp. pp. 353–358.

but can dispense entirely with the PEX and PMAEX. But do we really have any right to assume that all valid forms of inference are indispensable to or even useful in science (and mathematics)? The PEX and PMAEX are principles whose primary application and usefulness is within the area of religious and metaphysical thought. But why should there not be valid epistemic principles that are of little or no use to scientists, but which are useful, or even indispensable, to certain forms of religious and metaphysical thinking?

The PEX has, furthermore, a great intuitive appeal. If one comes out of a dark house and into the sunlight on a beautiful summer's day, the force and beauty of the day, of the trees, the sky and the flowers, may just strike one as intensely remarkable; and if they do, one may think softly to oneself that there must be (or may well be) something behind all this, something we don't know, some extraordinary or remarkable power or being we cannot fathom. I have had this sort of experience, and so, I think, have many other people. And for those who have had such an experience—either with respect to the beauty of a summer's day, the starry heavens, or the phenomenon of a new-born baby—the PEX and the PMAEX will have an immediate intuitive appeal, will make obvious sense. Of course, there are many people who have never seen or felt the remarkableness of the world, or of a summer's day, in any intense way, but this fact clearly does not by itself threaten the validity of the PEX. And there are others who have felt in this intense way about the world in one or another of its aspects, but who have not thought, as a result, that there had to be something remarkable behind things. But the fact that some people who feel the extraordinariness of things do not go on to posit the existence of something extraordinary behind them also does not vitiate the PEX. For the PEX says only that if something is found to be remarkable, one has *some reason* (and it is reasonable *other things being equal*) to think that it has been brought about by something remarkable. When a scientist refuses to postulate something extraordinary behind things he deeply feels are extraordinary, it is, I wish to claim, because for him other things are *not* equal. His reason to posit an extraordinary explanans is overridden by other rational considerations of a kind I shall be discussing just below. I think, then, that when we understand the PEX in the weak sort of way in which it is stated, the reluctance of scientists and others to postulate extraordinary explanantia does not vitiate the PEX, but can, rather, be best accounted for by assuming that such reluctance is the result of the fact that scientists and others allow a certain methodological principle of scientific inquiry, which they accept, to override the PEX, which they also accept.[17]

Some scientists and scientifically-minded philosophers, for example, Einstein,[18] come to believe in something extraordinary behind things on the basis of the awesomeness or extraordinariness of things they find in nature. Others, like J. J. C. Smart,[19] admit the awesomeness, etc., of certain things, but do not choose to postulate anything remarkable behind them. The reason why this occurs, I wish to suggest, is that such men accept both the PEX and another principle that often conflicts with the PEX, and in weighing the one principle against the other, some give greater weight to the PEX and others to the principle that opposes it. But what is this principle that opposes the PEX?

I think one of the reasons why a scientist is loathe to postulate a God or some other extraordinary force or being to explain things he finds extraordinary is his reluctance to posit things that do not fit neatly into the scientific theories he already accepts. Science typically seeks not only a greater and greater accumulation of knowledge, but also more and more systematic knowledge. As R. Rudner puts it: "System is no mere adornment of science, it is its very heart. . . . It is an ideal of science to give an organized account of the universe. . . ."[20] Now to posit a deity or some other kind of extraordinary force behind things is to posit an entity that does not fit well into the framework of present scientific theories, an entity whose existence and behavior are not readily understandable within present systematic scientific

[17] I should like to add that I do not wish to make this paper depend entirely on the validity of the PEX, of which I myself am not entirely convinced. But the PEX is important even if it is not valid, because it is a principle that seems to be presupposed in certain important areas of human thought. Even if the PEX is not valid, it is worth asking why people make use of it. And even if the PEX reflects a basically primitive pattern of thought, it will be worth noting the various ways that principle plays a part in areas of thought that are frequently considered not to be primitive.

[18] In his "Autobiography" in *Albert Einstein: Philosopher-Scientist*, ed. by P. A. Schilpp (New York, 1951), p. 9, Einstein says: "[The development of the world of thought] is in a certain sense a continuous flight from 'wonder'. A wonder of such nature I experienced as a child of 4 or 5 years, when my father showed me a compass. That this needle behaved in such a determined way did not at all fit into the nature of [familiar] events. . . . I can still remember—or at least believe I can remember—that this experience made a deep and lasting impression on me. Something deeply hidden had to be behind things."

[19] See "The Existence of God," *New Essays in Philosophical Theology, op. cit.*, p. 46.

[20] "An Introduction to Simplicity," *Philosophy of Science*, vol. 28 (1961), p. 112. Also see Sir Arthur Stanley Eddington's *The Philosophy of Physical Science* (Cambridge, 1949), p. 45; and Kant's *Critique of Pure Reason*, ed. by Norman Kemp Smith (London, 1963), B673–678.

thought, an entity which may operate by different laws from those currently known and whose relation to other entities presently posited by scientists is not currently understood or likely soon to be understood. Thus, for a scientist to posit a mysterious extraordinary being as the ground of (some aspect of) the (physical) world is in effect for him to thwart the inherently scientific aim of systematic unity within scientific theory and scientific knowledge, i.e., to accept at least a temporary lessening of the systematic unity of knowledge about the world that it is the very nature of science to seek. I think the following *Principle of the Systematic Unity of Science* is a valid epistemic principle governing scientific inquiry:

> It is, other things being equal, unreasonable (and there is some reason not) to explain phenomena in terms of entities the positing of whose existence decreases or makes likely a decrease in the systematization or systematic unity of (our body of) scientific knowledge.

It should be clear how this principle can come into conflict with the PEX, for in cases where something we have observed seems extraordinary, the PEX gives us some reason to think that some extraordinary being is responsible for it, while the Principle of the Systematic Unity of Science (henceforth, PSUS) gives us reason to think that we should not posit such an entity. The situation with respect to these two principles and extraordinary phenomena is indeed very much like the situation with respect to moral principles that conflict with respect to a given act. A man may adhere to the (*prima facie*) moral principle that the fact that an act involves helping someone get what he wants gives one some reason (makes it, *ceteris paribus*, reasonable) to think that it is morally right, and also to the (*prima facie*) moral principle that the fact that an act involves lying gives one some reason (makes it, *ceteris paribus*, reasonable) to think that it is morally wrong. And when a given act involves both lying and helping someone else—lying in order to spare someone's feelings, for example or lying in order to prevent a would-be murderer from finding his victim—one may find it hard to resolve the conflict between those two principles, i.e., to decide whether the act in question is right or wrong. Similarly, when confronted with a situation where the PEX and the PSUS conflict, it may be hard to decide which principle to give greater weight to, and thus to decide whether to posit an extraordinary entity and go against systematic unity or not to posit that entity and preserve that unity. Thus a philosopher like Kant might always give greater weight to the moral principle about lying, whereas most people would say that sometimes the principle of helping people overrides the principle of not lying,

RELIGION, SCIENCE, AND THE EXTRAORDINARY 199

so that it is sometimes right to help someone in a given case get what he wants even if it involves lying. And similarly, someone like Einstein may feel so strongly about the PEX that he finds it reasonable to posit something extraordinary behind things he finds extraordinary, even at the expense of scientific unity, while many other scientific philosophers and scientists may find it more reasonable not to posit such an entity, but to preserve scientific unity instead.

It is an interesting question when it is reasonable to decide in favor of a particular one of a pair of conflicting moral or epistemic principles. Of course, where something is not considered remarkable, conflict between the PEX and PSUS is avoided because of the inapplicability of the PEX. And for primitive people who lack a well-developed and unified body of scientific knowledge the PSUS has little, if any, relevance, just because positing a God beating a drum or striking flints does not effect any particular desystematization of unified systematic scientific knowledge that they possess. This is *part* of the explanation of why primitives who posit extraordinary thunder-gods are more rationally justified in doing so than we would be, if we, with our accumulated and systematic scientific knowledge, were to posit such a being.[21] There is also a tendency for natural phenomena like thunder to seem at least slightly less remarkable and awe-inspiring after they have been scientifically studied.[22] And this is perhaps another reason why "modern man" has less reason (via the PEX) to posit something extraordinary to explain thunder, than a primitive man has. Similarly, an ancient who after taking an hallucinogen has an extraordinary dream or other experience has perhaps more reason to think that something extraordinary and wondrous has happened to him than someone who has the same sort of experience, but who thoroughly understands how hallucinogens work, and so the former will have more reason to posit an extraordinary cause of his experience than the latter.[23]

[21] Of course, the primitive may *not* be justified in positing the *particular* kind of extraordinary being he does, but only in positing *some* kind of extraordinary being. It is interesting to note, however, that if the PEX is valid, the primitive mythological view of the world is not as purely fanciful and devoid of rationality as many philosophers and anthropologists have thought. The positing of extraordinary beings to explain extraordinary phenomena is best conceived, perhaps, as the first gropings of rationality, rather than as prerational and purely mythical.

[22] See Heschel, *Man is Not Alone, op. cit.*, p. 37.

[23] Perhaps this is why learning that a given mystical or dream experience was partly due to neurosis, hypnotism, or hallucinogens makes us less inclined to feel that that experience supports the truth of religion. (If people felt that every religious experience ever had been thus induced, their faith in God and their religiosity in general would be seriously shaken.) For learning this gives us some reason to think and indeed tends to make us think of the experience as less

However, even given scientific knowledge of the causes and mechanisms of thunder, one might still find that phenomenon extraordinary. For although thunder may not seem so extraordinary *in the light of* (our understanding of) those things that bring it about, we might still find the whole phenomenon of thunder-taken-together-with-its-causes-and-mechanisms to be remarkable and amazing indeed. One might, that is, still find it amazing that there should be such a phenomenon as thunder brought about in the way it is and operating as it does, and be thereby tempted to posit some underlying extraordinary explanation of that *whole* phenomenon.[24] Similarly with other natural phenomena. Even one well versed in science may still find it amazing that anything (physical) exists at all,[25] or that nature is as inexorably lawful as it is.[26] One may find it miraculous that there are no miracles.

Even if scientists and scientifically informed people will consider many things extraordinary or amazing, they will, presumably, also be less inclined than other people to use the PEX, because they are inclined to give greater weight than other people to the PSUS. For a scientist's major goals and interests are usually scientific, and so he is likely to give greater weight to the PSUS than others who lack those goals and interests, just because by sticking to the PSUS he tends to further the fundamental goal science has of explaining phenomena in a systematic, unified way. But the fact that something goes against the purposes of science does not necessarily mean that it goes against reason itself, and can in no way be rationally justified. And so a person with less commitment to science or with great aesthetic and religious sensitivity and capacity for wonder and awe might not be entirely irrational to give greater weight to the PEX than to the PSUS, and so come to believe in a higher being after exposure to something amazing.

---

amazing and remarkable than we otherwise would. Those who say that the "fact" that those who have come to believe in God through special experiences were all neurotic does not show that the God they came to believe in does not *exist* are, strictly speaking, right. But they are wrong if they think that learning about these neuroses does not diminish our *reason* for believing in God, *if* we ever based our religious faith even partially on the fact that saints and others have had special religious experiences. For once we learn that the people who had these experiences were all highly neurotic, we can no longer so confidently use the PEX to argue (implicitly or otherwise) from the existence of such experiences to the (reasonableness of believing in the) existence of some extraordinary being behind them.

[24] See Heschel, *Man is Not Alone, op. cit.*, p. 30; and *God in Search of Man, op. cit.*, p. 45.

[25] See Smart, *op. cit.*, p. 46.

[26] See Konrad Lorenz, "On Aggression," *Encounter*, vol. 27 (1966), pp. 29ff.

In fact, one of the results of the present paper is that we have provided the foundation for a new sort of argument (new, at least, in philosophical or theological literature) for the existence of some sort of deity. It is an argument founded on the PEX, and it has a greater chance of providing reason to believe in God than any of the traditional proofs of God's existence, which have all by now been pretty much discredited.[27] The argument goes:

(a) The order and beauty of the universe (or the fact that there exists something physical, rather than nothing [physical] at all) is extraordinary, remarkable, etc.

(b) The PEX is valid.

Thus:

(c) There is *some reason* to believe that there exists an extraordinary force, being, or thing that created the order and beauty of the world (or that brought it about that there is something physical).

Two further steps are possible:

(d) The order and beauty of the world (or the fact that there is something physical) is so remarkable, etc., that we should disregard the fact that positing an extraordinary, etc., being or thing goes against the PSUS, and should give greater weight to the PEX than to the PSUS.

Thus:

(e) It is *reasonable* for us to believe that some sort of extraordinary force, being, or thing is responsible for the order and beauty of the world (or for the fact that there exists something physical).

This argument differs from most traditional arguments for God in that it does not attempt to prove the existence of anything, but only that we have reason, or that it is reasonable, to believe in the existence of something. Secondly, it does not argue for the existence of anything like the traditional all-good, all-knowing Judeo–Christian God. However, the being or thing whose existence it refers to is something very like a God, and is the kind of "God" most negative theologians want to restrict themselves to talking about. Thirdly, this argument is obviously really two arguments: one from the order and beauty of

[27] If anyone thinks the traditional proofs have any life in them, he should read Alvin Plantinga's *God and Other Minds* (Ithaca, 1967), *passim*.

the world, the other from the fact of contingent physical existence. And each of these arguments runs parallel to one of the traditional arguments for God's existence. The traditional Cosmological Argument for God starts with the fact that there is some contingent (physical or mental) entity and tries to prove the existence of God as "First Cause." We have provided an argument that starts from the fact that there exists some contingent physical entity and tries to prove that there is reason (and that it is reasonable) to believe in something like a God. We have also provided an argument for some sort of God from the fact of the order and beauty of the world. And this argument parallels the traditional Argument from Design, inasmuch as both arguments start with the order and beauty of the world and from this try to prove something about the existence of something like a God.[28] Indeed with respect to both the arguments presented here, it might be said that what we have done is to take facts that have frequently been adduced as reasons for believing in God's existence and show for the first time just *how* those facts support the existence of (some sort of) God.

The two arguments we have introduced are not beyond criticism. One may question whether the order and beauty of the universe are really so remarkable (or, indeed, whether the world really is so orderly and beautiful). One may question whether it is extraordinary and amazing that something physical exists. One may question the PEX or assumption (d). But in any case the arguments have, I think, a certain plausibility to them. (Indeed to many ordinary people they will not appear particularly striking or new, just because they are so pervasive implicitly in so much ordinary religious thinking.) Of course, many philosophers and scientists will not be very happy with the above arguments, especially with the move from (c) to (e). Nor am I happy myself with that move. And this is in great part, I think, because of our commitment to scientific inquiry. Most scientifically minded people, and I myself, are inclined to stress the PSUS over the PEX and not to assent to premiss (d) of the above argument(s). However, I can see no way of *showing* that (d) is an unreasonable premiss, nor thus that the argument for (e) cannot be made to go through. This seems to be a topic that deserves further serious attention.

[28] In his introductory remarks about the Argument from Design (what he calls the Physico-theological Proof) in the *Critique of Pure Reason* (B650), Kant seems to have seen or to have been close to seeing the possibility of an argument from the extraordinariness of certain aspects of the world to the existence of some sort of God. But he does not seem to see the argument from extraordinariness as an argument separate and distinct from the traditional Argument from Design itself.

IV

We have understood certain problems about the reasonableness of belief in some sort of deity in terms of a conflict between the PEX and PSUS. But there are yet other intellectual situations that can be understood in terms of a conflict between these two principles. Consider, for example, the intellectual climate that exists today about the so-called phenomena of ESP. Some people believe in ESP and others do not; but there are, I think, three major stances that can be taken with respect to ESP, and these can be clarified in terms of the PEX and the PSUS. Some psychologists and men of science think that putative ESP phenomena can be explained away as being due to cheating or as merely chance phenomena. With regard to the ESP card-guessing experiments, on which the proponents of ESP place so much weight, it is often claimed that peeking may have occurred, or else that there are purely statistical reasons to expect at least some subjects to perform as well as those few whose performances at card-guessing have been used to support the existence of ESP. Such psychologists and men of science do not, therefore, find the results of ESP experiments particularly remarkable or extraordinary, and so have no motive from the PEX to explain the results of ESP experiments in terms of some extraordinary power of Extrasensory Perception.[29]

Others do not think that the results of ESP experiments can be so easily explained away. They think that something quite remarkable is taking place, but prefer not to jump immediately to the conclusion that there exists a force or power of ESP, out of a desire to preserve as much as possible the present order and unity of our scientific understanding of the world. For to postulate ESP would be to posit a power or force that cannot be readily understood within the scope of present scientific theories and that is totally different from anything currently accepted in science; nor does there seem to be any way, at present, of incorporating both ESP and the phenomena science already reckons with into some new coherent unified body of beliefs. If one really accepts ESP, one in effect accepts a lessening of the systematic unity of our scientific knowledge about things, at least for the present. Thus many scientists who find the results of ESP experiments remarkable in themselves refuse to posit ESP, because

---

[29] It should be clear that ESP is supposed to be extraordinary even by its proponents, since only a few people at very rare moments are supposed to possess it, and it enables them to do things people cannot ordinarily do. ESP is extraordinary and remarkable in something like the way Superman's powers are supposed to be extraordinary and remarkable.

of their allegiance to the PSUS. Instead, they attempt, or think one should attempt, to understand those results in terms of some perhaps strange or unexpected interaction of forces or entities that are already posited and fairly well understood by scientists. Such people, then, give greater weight to the PSUS than to the PEX. Others, of course, believe in ESP, and they can perhaps best be understood as people who find the results of ESP experiments remarkable and whose nonallegiance to the PSUS (or whose amazement at the results of ESP experiments) is great enough that they will give greater weight to the PEX than to the PSUS, and so will posit ESP as the explanation of the results of those experiments.[30]

I am inclined to think it more reasonable for most people at the present time not to believe in ESP than to believe in ESP firstly because of the chance of foul play by ESP experimenters, and secondly because I do not think one should rush in and posit an extraordinary explanans for every new extraordinary thing that one observes or hears about. Respect for the "old truths" and for the preservation of previously won scientific *systems* makes it reasonable, I think, to refrain from positing something (like ESP) that does not fit in well within the current organized body of scientific knowledge, and first to make every attempt to fit the remarkable phenomenon in question into already existing patterns of scientific explanation, to explain it, that is, in terms of some interrelation of things, forces, and laws already posited and investigated.[31] However, if such attempts persistently fail, then perhaps the PSUS should give way and one should posit an extraordinary explanans. At present, I think, we have not done all we can or should to explain the results of ESP experiments in terms of existing theories and ideas. So I think positing ESP is premature. That is, in present circumstances we ought to give greater weight to the PSUS than to the PEX with respect to ESP experiment results. But this may not always be so in the future.

Whether or not it is reasonable to believe in ESP at present, the current existence of disagreements among scientists and others with respect to putative ESP phenomena can be explained in terms of our

---

[30] Unless they are really the charlatans some people say they are, Rhine and Soal, who have conducted ESP experiments, belong in this last class. One can perhaps explain their readiness to believe in ESP as being a result of their having had the results of the ESP experiments at first hand and thus being more amazed by those results than those who only know of them at second hand. Note further that one reason why ordinary people are generally more willing to accept ESP than scientists are is that ordinary people are generally less committed than scientists to the goals of science, and thus to the PSUS.

[31] As is suggested by Michael Scriven in "Modern Experiments in Telepathy," *The Philosophical Review*, vol. 65 (1956), p. 249.

two principles, the PEX and the PSUS. And these principles can be used to explain other scientific or intellectual disagreements as well. Current disagreements about flying saucers or ghosts can, I think, be understood in much the same way as we explained current disagreements about ESP. And inasmuch as many phenomena of intellectual life can be plausibly understood in terms of our two principles, we can see the prevalence and pervasiveness of those two principles in human thought, and this in turn is some sort of reason additional to those offered earlier for thinking those principles valid, and for explaining certain conflicts within the religious sphere as resulting from a conflict between them.

In recent decades, religious thinkers and philosophers have often held that if religious belief and thought are not rational by scientific standards, have no justification *of the sort* that scientific theories have, they are not rational at all. And so they have thought that religious belief had to be either scientifically verified or else some cognitively meaningless emotive way of seeing and responding to things. Since there does not seem to be any available way to verify religious beliefs in the manner of scientific hypotheses, they have tended to conclude that religious belief is irrational, and blind, or, that religious "belief" is really just an attitude, emotion, or "blik"[32] to which rationality is totally irrelevant—neither of which is a very pleasant prospect for religionists to have to face. If the theory of the present paper is correct, however, there is a middle way out of this predicament, and there is more to religious belief (in the existence of God, at least) than sheer emotionalism and/or irrationality. There are some rational grounds for religious belief, but these are not essentially *scientific*. Rather, there is a rationality *sui generis* to religious thought that is not encompassed by science, or by purely scientific standards of the reasonable acceptance of hypotheses: a rationality whose nature and significance are only just beginning to be understood.

Columbia University

---

[32] For clarification of the notion of a "blik," see "Theology and Falsification" (esp. section by R. M. Hare) in Flew and MacIntyre, *op. cit.*

Index of Names

Achinstein, Peter 9–29, 111, 126, 138–40, 183, 186
Ackermann, Robert 118, 126
Agassi, Joseph 162–70
Altham, J. J. 140
Aristotle 56

Bar-Hillel, Yehoushua 170
Barker, Stephen F. 111, 126, 138–40
Bauer, Edmond 91
Blackburn, Simon 128–42
Bohm, David 89
Bohr, Niels 62, 71, 72, 79, 83, 86
Boltzmann, Ludwig 71
Born, Max 77, 79
Bourbaki, Nicolas 63
Boyle, Charles 9
Bradley, Francis Herbert 15, 20
Bridgman, Percy W. 71
de Broglie, Louis 71
Bromberger, Sylvain 15–17, 18
Bub, J. 89
Bunge, Mario 61–99

Carnap, Rudolph 35, 143, 150–52, 157–8, 159, 160, 165, 166, 168, 169, 170, 171, 186
Chisholm, Roderick M. 35
Copi, Irving 148
Craig, William 32

Daneri, A. 89
Darwin, Charles 163
Dewey, John 71, 72
Dingler, Herbert 71
Doyle, Arthur Conan 173
Duhem, Pierre 51

Eberle, Rolf 30, 37
Eddington, Arthur 58, 71, 171, 172, 175, 182, 186, 197

Einstein, Albert 44, 57, 71, 163, 197, 199

Fain, Haskell 117, 118, 126
Feyerabend, Paul 59, 79, 171, 175, 183, 184, 186
Fine, Arthur 59, 184, 186

Galileo Galilei 183
Goodman, Nelson 100–26, 128–42, 195
Groenewold, H. J. 89
Grünbaum, Adolf 47, 52, 57, 58
Grunstra, Bernard R. 100–27

Hacking, Ian 138, 140
Hanen, Marsha 102, 126
Hare, R. M. 205
Havas, Peter 56
Heisenberg, Werner 72, 79
Hempel, Carl G. 30, 31–3, 35, 40, 102, 111, 126, 160
Heschel, Abraham 190, 191, 199, 200
Hesse, Mary B. 30, 184, 186
Hilpinen, Risto 36
Hintikka, Jaakko 36
Hoyle, Frederick 181
Hullett, James 101, 103, 104, 117, 127
Husserl, Edmund 91

Jeffrey, Richard C. 35
Jevons, F. B. 188

Kant, Immanuel 56, 72, 99, 122, 198, 202
Kemeny, John G. 157, 158, 160
Kennedy, John F. 21
Kepler, Johannes 183
Kerr, Clark 174, 186
Kuhn, T. S. 183, 186
Kyburg, H. E. 30, 35, 137–8

INDEX OF NAMES

Lamarck, Jean Baptiste 163
Leblanc, Hugues 101, 127
Lehrer, Keith 30–41
Levi, Isaac 36
Lincoln, Abraham 21
Lindholm, Lynn 162
Loinger, A. 89
London, Fritz 91
Lorenz, Konrad 200
Ludwig, Günther 86

Mach, Ernst 71, 72, 185
Margenau, Henry 89
Markov, A. A. 83
Massey, Gerald 44
Maxwell, Grover 171, 175, 181, 186
Mellor, D. H. 171–87
Mill, John Stuart 180
von Mises, Richard 82

Nagel, Ernest 180, 184, 186
von Neumann, John 79, 87–90, 91, 98
Newton, Isaac 58, 163, 183
Neyman, Jerzy 143
Nowell-Smith, Patrick 192

Oppenheim, Paul 160

Pailthorp, Charles 148
Peirce, Charles Saunders 71
Planck, Max 71
Plantinga, Alvin 201
Poincaré, Henri 43, 44, 45, 46, 47, 49–51, 56, 57
Popper, Karl R. 83, 162–70, 183, 186
Prosperi, G. M. 89
Putnam, Hilary 57, 59

Quine, W. V. 177, 179, 181, 186

Reichenbach, Hans 43, 44, 45, 51, 52, 54, 57, 60, 82
Rhine, J. B. 204
Riemann, Bernard 47
Rosenfeld, L. 86
Rudner, Richard 197
Russell, Bertrand 180, 186
Ryle, Gilbert 171, 173, 175–79, 186

Savage, L. J. 143
Scheffler, Israel 102, 127
Scheibe, Erhard 96
Schilpp, P. A. 150, 152
Schlesinger, George 185, 186
Schopenhauer, Arthur 189
Schrödinger, Ernst 71, 78, 87, 89
Schwartz, Robert 101, 103, 104, 117, 126, 127
Scriven, Michael 34, 204
Seid, Robert 146
Sellars, Wilfrid 41, 171, 175, 186
Silberblatt, S. 190
Sklar, Lawrence 42–60
Skyrms, Brian 101, 111, 127
Slote, Michael Anthony 188–205
Smart, J. J. C. 197, 200
Soal, S. G. 204
Spielman, Stephen 143–61
Stebbing, Susan 172, 186
Stevens, S. S. 74
Strawson, P. F. 171, 177, 187

Thomson, Judith Jarvis 118, 127

Wigner, E. P. 62
Wisdom, John 192
Wittgenstein, Ludwig 141, 168